Biochimica et Biophysica Acta:

The Story of a Biochemical Journal

Biochimica et Biophysica Acta:

The Story of a Biochemical Journal

E.C. SLATER

1986

ELSEVIER

Amsterdam · New York · Oxford

© 1986, Elsevier Science Publishers B.V. (Biomedical Division)

All rights reserved. No part of this publication may be reproduced, stored in a retrieval system or transmitted in any form or by any means, electronic, mechanical, photocopying, recording or otherwise without the prior written permission of the publisher, Elsevier Science Publishers B.V. (Biomedical Division), P.O. Box 1527, 1000 BM Amsterdam, The Netherlands.

Special regulations for readers in the USA:
This publication has been registered with the Copyright Clearance Center Inc. (CCC), Salem, Massachusetts.
Information can be obtained from the CCC about conditions under which the photocopying of parts of this publication may be made in the USA. All other copyright questions, including photocopying outside of the USA, should be referred to the publisher.

ISBN: 0-444-80769-1

PUBLISHED BY
Elsevier Science Publishers B.V. (Biomedical Division)
P.O. Box 211
1000 AE Amsterdam
The Netherlands

SOLE DISTRIBUTORS FOR THE U.S.A. AND CANADA:
Elsevier Science Publishing Company, Inc.
52 Vanderbilt Avenue
New York, NY 10017
USA

Library of Congress Cataloging in Publication Data

Slater, E. C. (Edward Charles), 1917-
    Biochimica et biophysica acta.

    Includes bibliographies.
    1. Biochimica et biophysica acta–History.
2. Biological chemistry–Periodicals–Publishing–
Netherlands–History.   3. Biophysics–Periodicals–
Publishing–Netherlands–History.   I. Title.
QP511.S57   1986   574.19′2′05   86-8975
ISBN 0-444-80769-1

Printed in The Netherlands

Preface

APOLOGIA

One of the first things that I learned about preparing a scientific paper is to write the INTRODUCTION last of all, since in the course of assembling the RESULTS and making profound statements in the DISCUSSION, the paper develops an identity of its own that might differ quite considerably from what the author had in mind when he began writing. It is certainly the case with this book.

As the time approached when I would have to give up a 'direct' involvement in biochemical research (that is to say supervision of students) with, as I firmly believe, the consequence that I should retire from my editorial functions, the idea developed of writing a history of BBA as my swan song. I was encouraged in this direction by Elsevier and, originally, joint authorship with Dr. Jeff Hillier, with contributions from the Editorial Secretariat, was envisaged. However, as I delved more deeply into the archives and the book began to take shape, it became clear that, as in another activity that

I indulge in, I would feel more free if it were a single-handed effort. That is the first of my apologies.

The second apology also concerns the egoistical character of the book. It is related to the nature of human memory when the human being in question (this one at least) is approaching his three score years and ten. I have a vivid and mostly accurate memory of events more than thirty or even twenty years ago, and do not have many problems about the last five years except for a compressed time scale, but while reading the letters written and the minutes taken in the intervening period, I found that I had completely forgotten many events. Thus, despite my involvement at the time, I found myself often thinking of the Slater named as another person. That is one of the reasons why I refer to myself by name in this book. A second reason is that I am embarrassed by using the first person so frequently, although it is scarcely logical to regard the equally frequent use of one's surname as less egoistic. The third reason is that I hoped that avoidance of the first person would help me subconsciously to be as objective as possible.

Running a journal like BBA, with a turnover of more than 5 million dollars, is a joint venture requiring close cooperation between Publisher, Managing Editors and Editorial Secretariat. However, the interests and motivation of the three parties are not identical and tact and a willingness to compromise are required of all parties. When there are good personal relations, everything runs smoothly most of the time, but situations can arise involving important questions of principle when even the best of personal relations cannot prevent tensions developing. Just as a sea voyage without storms makes dull reading, conflicts are more interesting for the reader than the long periods of smooth relations. However, concentrating on these tensions can give an unbalanced picture of the true relations between Publisher, Managing Editors and Editorial Secretariat. For this, I offer my third apology.

My fourth is for the fact that, from reading this book, one might imagine that only the three parties mentioned in the above paragraph are involved in producing BBA. This is of course not true.

The most important omission is the author, without whom there would be no BBA, but who gets short shrift in this book. I had hoped to add a chapter in which some of the interesting and illustrative correspondence with authors was included. A number of practical difficulties, the most important of which was insufficient time before my scientific archives were moved to the Provincial Archives of North Holland, prevented this. Respect for the privileged character of such correspondence had also to be kept in mind. During the International Congress of Biochemistry in Amsterdam last summer, I was told of a case which I had completely forgotten where we had made a ghastly mistake in rejecting a paper on the basis of what turned out to be an unfair and even not disinterested referee's report. However, my informant mentioned this not to remind me of the mistake but of the way we handled it when the true facts were drawn to my attention. We followed Westenbrink's dictum of freely and unreservedly admitting our mistake and giving the paper accelerated publication to make up for lost time. It now transpires that by so doing we kept a young author in science. This is a success story, but how often have there been cases when the wrong decision has had unhappy consequences? Like the surgeon, we would never sleep if we dwelled on this. We can only do our best to keep failures to a minimum. The literally thousands of editors and referees who do the real donkey work of evaluating papers and are responsible for our correct decisions also receive far too little attention in this book.

Finally the desk editors or Editorial Office, who are the link between the Editorial Secretariat and the printers, are scarcely mentioned except in Fig. 10. The reason for this is that this office works directly under the responsibility of the Publisher in contradistinction to the Editorial Secretariat who are directly responsible to the Managing Editors. However, the Head of the Editorial Office and often a member of his staff do attend the regular meetings of Managing Editors, Publisher and Editorial Secretariat and make an important contribution to decisions of a technical nature, such as the all-important question of publication time.

ACKNOWLEDGEMENTS

If I were to acknowledge all those who have helped me in my function as Managing Editor of BBA, this Preface would become unreadably long. I shall confine myself then to those who have directly helped with this book, with one exception – Professor Westenbrink. I had a great admiration for Westenbrink as a man and for his achievement in founding BBA and I have tried to follow his example.

In the early chapters, I have drawn heavily on reminiscences of the earlier editors published in the Westenbrink memorial in BBA, especially the admirable obituary by Professor Max Gruber. I am also deeply indebted to the co-founder of BBA – Mr P. Bergmans – for the notes that he provided, within a few days of the request to do so, of BBA's early years as seen by the Publisher. Mr. Jan Geelen was not quite so quick, but I am grateful for his reminiscences of the founding of the Editorial Secretariat, which was his idea, his revelations of the Publisher's initial reaction to the proposal and of the admonitions he received both from the Publisher and myself of what he should not do! These are included as an Addendum to Chapter 4.

A special acknowledgment is due to the Publisher, who has not objected to my sometimes critical comments. I have tried to present the arguments and disagreements objectively. In one or two cases – too few I am sure – I have admitted directly or implied that I was wrong. If I imply in other cases that I still think that I was right, I hope that I have presented the case for the Publisher sufficiently fairly for the reader to make up his own mind.

I would like to thank Professor Laurens van Deenen, Chairman of the Board of Managing Editors, and Dr. Jeff Hillier for reading the manuscript and for their many constructive suggestions, most of which I have incorporated. However, the responsibility for the final text is mine alone. I also wish to acknowledge with thanks help from the Editorial Secretariat and Mr. Ralph Lupton, formerly Head of the Editorial Office of BBA, and Mr. Gerald Mettam for updating certain data and for drafting the figures.

I commend to you especially the last chapter – *1986 and beyond* – written by Dr. Hillier, who is now responsible for the financial and policy management of the journal for the publisher. This thoughtful and informative discussion of the future of scientific publishing could never have been written by an Editor, who is presumably an expert in his field of science but essentially an amateur where the technicalities of publishing are concerned. Dr. Hillier has provided the professional and expert approach that the subject requires. It is an excellent example of the scientific contribution from the side of the Publisher and the advantages to be gained by the harmonious working together of the biochemist with the expertise and financial resources of a large publishing house. Indeed, it seems likely that the maximal utilization of possibilities opened up by the new technical developments will require close cooperation of all concerned with biochemical publishing, including the biochemical societies. The IUB Commission of Editors of Biochemical Journals could provide a forum for this purpose.

E.C. Slater

Department of Biochemistry
University of Southampton
England
April, 1986

Contents

Preface v
    Apologia v
    Acknowledgements viii

Chapter 1 The birth of BBA 1
    Biochemical journals between the wars 1
    Enzymologia 3
    H.G.K. Westenbrink (1901-1964) 8
    Elsevier Publishing Company 14

Chapter 2 The first volume 19

Chapter 3 The early years 27
    Financial problems 27
    Short Communications and Preliminary Notes 29

Chapter 4 The second decade 39
    Slater joins Westenbrink 39
    The 'third' man 44
    Replacement for Westenbrink 46

	Information exchange groups	50
	Quality of papers	52
	Notes to contributors	54
	Sectionalization of BBA	57
	Appendix – The birth of the BBA Secretariat (by J. Geelen)	59
Chapter 5	**The growth explosion of BBA in the 1960's**	63
	The 1966 'Statement of Policy'	63
	Failure to contain the growth of BBA	71
	Data deposition	75
Chapter 6	**The problems of the 1970's**	77
	Number of subscribers and the price of BBA	77
	The Editorial Secretariat	83
	Relations between Managing Editors and Publishers	87
	BBA Reviews	97
Chapter 7	**BBA enters the 1980's and makes ready for the 1990's**	99
	Restructuring the Board of Managing Editors	99
	Restructuring of BBA sections	102
	Publication time	103
	Language of BBA	104
	Computerization	105
	Decentralization	106
	The crystal ball	111
Chapter 8	**1986 and beyond (by J. Hillier)**	113
	Technology and the author	115
	The peer review system	117
	Editorial and production	119
	Distribution and ordering	120
	The user	121

CHAPTER 1

The birth of BBA

BIOCHEMICAL JOURNALS BETWEEN THE WARS

Those of us who have lived through the period of great scientific expansion since 1945 tend to forget that biochemistry also developed rapidly after the first World War, especially in Europe. Most of the results of this research were published in a small number of journals: *Hoppe Seyler's Zeitschrift für Physiologische Chemie* and *Biochemische Zeitschrift* in Germany, the *Biochemical Journal* in England, the *Bulletin de la Société de Chimie Biologique* in France and the *Journal of Biological Chemistry* in the United States of America. There were at that time few specialist journals, with one notable exception to which we shall return. *Nature* also was, as it is now, an important vehicle for publication, especially in the form of 'Letters to the Editor', and some leading English biochemists, notably D. Keilin, published many of their now classical papers in the *Proceedings of the Royal Society*. Fig. 1 gives an idea of the development of the individual journals between 1920 and 1945. It should be noted, however, that since no allowance has been made for differences in

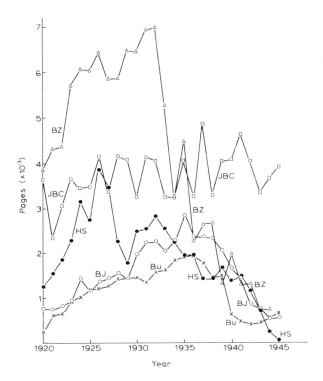

Figure 1. Size of the major biochemical journals between the Wars. ×, *Bulletin de la Société de Chimie Biologique* (Bu); ○, *Biochemical Journal* (BJ); ●, *Hoppe-Seyler's Zeitschrift für Physiologische Chemie* (HS); □, *Journal of Biological Chemistry* (JBC); △, *Biochemische Zeitschrift* (BZ).

the size of the page, the number of pages of the different journals does not accurately reflect the relative size of the journals.

The four European journals grew rapidly in the twenties, and the *Biochemical Journal* and the *Bulletin* continued to grow until 1935. After nearly trebling its size between 1920 and 1927, however, *Hoppe Seyler* slowly declined until in 1939 it was about the same as in the early twenties. *Biochemische Zeitschrift*, already a large journal in 1920, grew rapidly in the next 5 years and, after a slow further growth, declined precipitously in 1933, until in 1939 it was smaller than *Hoppe Seyler*. All four journals declined during the War, the two German journals virtually disappearing in the period 1940–5. The effects on the European victors were, however, scarcely less dramatic, so far as biochemical publication is concerned, the *Biochemical Journal* publishing in 1945 only about one-quarter the number of pages published in 1939. In contrast, the *Journal of Biological Chemistry* remained almost constant during the entire period 1920–1945.

It is noteworthy that the first peak in biochemical publication was reached long before the War. Surely the establishment of the Third Reich and the expulsion of Jewish biochemists from Germany contributed greatly to the decline of German biochemical publications. It should be remembered that in the 1920's German was still an important language of scientific communication and in many English-speaking countries courses in German reading were compulsory for the science student. The countries that belonged to the former Habsburg Monarchy and Scandinavia and the Low Countries were also still very much oriented towards Germany. It seems likely that the damage done to Germany's reputation as a cultured country by the rise of the Nazi party resulted in many non-Germans ceasing to publish in German journals. However, further research is necessary to establish this.

It is not immediately apparent why *Hoppe Seyler* should have peaked so much earlier than 1933, nor indeed why the *Biochemical Journal* and the *Bulletin* started their decline in 1935–1936. Possibly the financial crisis played a role.

ENZYMOLOGIA

The important exception to the statement above that there were few specialist journals is *Enzymologia*. This journal was founded in 1936 by Carl Oppenheimer (1874–1941), then already well known for his monumental treatise on *Fermente und ihr Wirkungen*

[1] as well as for his textbook *Grundriss der Biochemie* [2]. *Enzymologia* was published by Dr. W. Junk, who had fled Germany to The Hague. Oppenheimer, like Junk a Jew, but with no official or University position, remained living in Germany until 1938 when he too was driven into exile to Holland.

Enzymologia was immediately a success, and attracted many of the leading biochemists from all over the world. Volume 1 contains 44 papers (396 pages) written in English, French and German, beginning with one by the famous Dutch scientist A.J. Kluyver (Beziehungen zwischen den Stoffwechsel-Vorgängen von Hefen und Milchsäure-Bacterien und den Redox-Potential im Medium) and including five papers by Carl Neuberg, who had been Editor of *Biochemische Zeitschrift* until 1935. Other names that spring out from the contents page are: S.J. Folley and H.D. Kay, J.H. Quastel, K. Linderstrøm-Lang, J.B. Sumner, H. Tauber, R. Willstätter, K. Myrbäck, Z. Dische, P. Chaix and C. Fromageot, S. Adler, P. Fleury and J. Courtois.

Between 1938 and 1940, eight volumes were published, including two (Volumes 3 and 4) published in 1937 as a Festschrift on the occasion of Carl Neuberg's 60th birthday. In Volume 4, we find the first paper on the Krebs tricarboxylic acid cycle (The role of citric acid in intermediate metabolism in animal tissues) [3], which had been offered to *Enzymologia* after it had been refused by the Editor of *Nature*, because:

> "as he has already sufficient letters to fill the correspondence columns of *Nature* for seven or eight weeks, it is undesirable to accept further letters at the present time on account of the delay which must occur in their publication. If Mr. Krebs does not mind such delay, the Editor is prepared to keep the letter until the congestion is relieved in the hope of making use of it. He returns it now, however, in case Mr. Krebs prefers to submit it for early publication to another periodical." [Reproduced from [4] with permission.]

Hermann Kalcker published in Volumes 2 and 5 the first papers on oxidative phosphorylation in a cell-free system [5, 6].

A few days after the German invasion of Holland in May 1940, Oppenheimer asked H.G.K. Westenbrink, from the Laboratory of Physiological Chemistry of the University of Amsterdam, to join him on the editorial staff. Volume 8, published in 1940, still has Oppenheimer's name as Editor, but it was struck off Volume 9, which appears with Westenbrink as editor. Westenbrink asked K. Linderstrøm-Lang of the Carlsberg Laboratory, Copenhagen to join him as Editor. Oppenheimer continued to advise Westenbrink from behind the scenes, but he died in 1941. He had long been ill, but his death was the result of an accident. His wife died soon afterwards, so they were both spared the suffering that befell their people in all of Nazi-occupied Europe.

Volume 9, containing 394 pages, appeared during 1940 and 1941. Among the papers may be mentioned two from Westenbrink, with two fugitives, Max Gruber and H. Veldman, among the co-authors. H. Veldman died later in a concentration camp. Max Gruber, happily, survived the war and later became one of the Managing Editors of BBA. There were also 12 papers from Albert Szent-Gyorgi's Institute for Medical Chemistry at Szeged, Hungary. Volume 10 appeared during 1941 and 1942, and Volume 11 between 1943 and 1945, the last number being issued in September, 1945, after the liberation.

In a volume of *Enzymologia* in the library of the Laboratory of Biochemistry (previously called Physiological Chemistry) of the University of Amsterdam (where Westenbrink worked when he was Editor of *Enzymologia*), is a printed loose sheet in which Westenbrink gives a short account of the vicissitudes of *Enzymologia* during the war. He explains that

> "the process of 'aryanization' of Dr. Junk's publishing house was carried out very slowly, in about four years, passing several stages, and was therefore the cause of a perpetual state of uncertainty with many crises of a pecuniary character. The many delays in the appearance of the journal, which caused so much annoyance to both editors and contributors, were the result of this. If I have been able to fulfil my promise to Oppenheimer to keep *Enzymologia* alive during the

war, I have, in the first place, those biochemists to thank for it who remained faithful to the journal, notwithstanding the many deceptions they experienced during all those years" and "The issue which you hereby receive was ready for despatch in September 1944, when the Netherlands were so sorely afflicted by the operations of war. We hope to be able soon to forward the next number, which has already been set up, but could not yet be printed owing to the prevailing lack of electric power. Thereafter, publication of articles received recently will follow immediately."

The loose sheet is not dated, but from its context and an accompanying order form, it was probably sent with Issue No. 4 of Vol. 11, dated September, 15, 1945.

Between the two paragraphs quoted above is the following:

"*Enzymologia* now begins a new life. It will be published in future by the Elsevier Publishing Company of Amsterdam, London, New York, and Brussels, one of the oldest* and greatest of Dutch publishing houses. The directors have assured me that they will give all possible care to the authors' contributions, and, above all, ensure speedy publication of manuscripts."

Moreover, on the first page is the following statement from Elsevier Publishing Company, Inc. addressed to "the contributors, subscribers and readers of *Enzymologia*."

"In pursuance of the report made by Dr. H.G.K. Westenbrink, we have pleasure in informing you that, at the request of the Editors, our publishing office has decided to resume the publication of the aforementioned periodical founded by Professor Carl Oppenheimer."

It is now apparent that either this announcement was premature or it was overcome by events. Westenbrink's student and col-

---

* Someone, some time, has underlined this word in pencil on the copy found in the library with a question mark in the margin (see, further, p. 14).

league, Professor Elizabeth Steyn Parvé [7], wrote in an 'In Memorium' to Westenbrink that

> "After the war, it proved impossible to come to terms with the publishers about the manner in which the journal ought to be continued in the eyes of the editors (Oppenheimer had died in the meantime). So Lang and Westenbrink decided to found a new journal, forseeing that there would soon be a wide demand for more possibilities of publication. Most of the preliminary negotiations fell to Westenbrink, who had also sought contact with the Elsevier Publishing Company. Finally all arrangements were completed and he wrote me a letter proudly announcing the name of the new journal and those of the first editorial and advisory boards. 'Isn't this international and is there, except for mine, one name of lesser reputation among these'?"

In the same 'In Memorium', Carl Cori [8], one of the founder editors, wrote:

> "The principal facts relating to the origin of the journal are these. Drs. Westenbrink and Linderstrøm-Lang had been Editors of *Enzymologia,* a journal which belonged to a Jewish publishing house and therefore had great survival difficulties during the Nazi occupation of Holland. After World War II, Drs. Westenbrink and Linderstrøm-Lang reached the conclusion that a journal of larger scope, a first-class journal devoted to all of Biochemistry and Biophysics, was an essential necessity for Western Europe and as a result they accepted an invitation by the Elsevier Publishing House to be Editors of the new journal."

There is some disagreement in these accounts as to whether Elsevier or Westenbrink made the first approach. Professor Elizabeth Steyn Parvé believes that difficulties arose between Westenbrink and Junk concerning the perennial problem of the influence of publisher on editorial policy, and that Westenbrink and Linderstrøm-Lang then sought another publisher. Westenbrink sought the advice of Dr. W. Gaade, who had joined Elsevier on September 1, 1946. Mr. P. Bergmans, who entered the service of

Elsevier two months later as head of the commercial section, recalls that Mr. J.P. Klautz, then director of Elsevier, 'got wind of' Westenbrink's plans and made the approach. It is possible that an initial unofficial contact between Westenbrink and Gaade was followed up by an official contact between Klautz and Westenbrink. Both Professor Steyn Parvé and Mr. Bergmans recall that Westenbrink and Linderstrøm-Lang had also consulted Professor Claude Fromageot concerning the new journal.

In the event, *Enzymologia* continued to be published by W. Junk Publishing House in the Hague as a small specialist journal. In 1973, its name was changed to *Molecular and Cell Biology*.

H.G.K. WESTENBRINK (1901–1964)

[REPRINTED FROM THE OBITUARY BY M. GRUBER [9]]:

Hendrik Gerrit Koob Westenbrink was born on January 30, 1901 in Assen, in the Dutch northern province of Drenthe. He spent only a few years of his childhood there as his family moved to Winschoten, where he received his secondary education. His father was director of a 'kweekschool' where primary-school teachers are trained, and his grandfather had also been a teacher. He thus hails from a family of teachers and he himself also possessed to a large extent the qualities of a good teacher. He liked to lecture, without ever being pedantic, and especially liked to help younger people to grasp problems. In this period Westenbrink got some local, and outside, fame as football (soccer) player. In 1919, after having finished his secondary education and a short period in the army, he started his studies in chemistry at the University of Groningen. He had hesitated whether he should study chemistry or further develop his gifts in drawing and painting. His motives for not taking up art as a profession were based on his conviction that a mediocre chemist might be of some use to world and society, while a mediocre artist would better remain an amateur. He always kept his love for drawing and painting, and especially in his later period again took painting lessons in order to improve his technique.

In 1924 he finished his formal studies with the degree of 'doktoraal'. He must have been a quick student as five years at that time was well below the average. He started work on his thesis and received his Ph.D. in 1926 on a dissertation devoted to an X-ray analysis of rhombic and monoclinic vitriols and their mixed crystals. He had done this work under the supervision of Professor Jaeger, who then had the chair of inorganic chemistry at Groningen. Westenbrink was one of the first people in Holland to apply the relatively new method of X-ray diffraction to structural analysis. During this time he had been an assistant, and later first assistant, teaching young students the secrets of chemical analysis.

After his Ph.D. he accepted a position at the Department of Physiology in Groningen where, as successor to Hamburger, Buijtendijk had been appointed. He stayed there only for about 9 months and then became first assistant at the Laboratory of Physiology in the University of Amsterdam under van Rijnberk.

It must have been quite a decision for a young physical chemist and crystallographer to switch to biochemistry at that time, and especially in the Netherlands. While Holland had quite a tradition in physiology and also physiological chemistry in the 19th century (Mulder, Pekelharing), the interest of chemists had turned to the problems of physical chemistry, inspired by figures like van 't Hoff and others, and the organic chemists were less interested in natural products than in applying physical chemistry to organic reactions, as in aromatic substitution. During his first years in Amsterdam, Westenbrink thus had to teach biochemistry to himself, which he did, as he liked to tell, by having to lecture to medical students. He mentioned that at first it took hard work to remain abreast of them. In 1929, the University of Amsterdam instituted a separate chair and laboratory for physiological chemistry. B.C.P. Jansen was appointed to this chair. He had been the first, together with Donath, to isolate a vitamin, now called thiamin, in pure and crystalline form. Westenbrink transferred to the new department which was in the same building as physiology. His interest in thiamin probably started at that time, an interest which would never flaw, but only widened into an interest in general metabolism. While Jansen continued his work on the chemistry of thiamin, Westenbrink started work on its physiological and metabolic role. He was later appointed a 'privaat docent' and his public lecture in 1931 when accepting this position was on "older and newer views on (tissue) respiration". In 1934 he became 'conservator' at this laboratory. In 1936 he worked for a period in Oxford with Sir Rudolph Peters, and in 1938 for a year at the Carlsberg Laboratory. From this time dates his friendship with Linderstrøm-Lang which was of decisive importance for the foundation and growth of *Biochimica et Biophysica Acta.*

When the war broke out and the Netherlands were occupied Westenbrink tried to continue his work as long as possible, without making concessions to the German authorities. Moreover, he tried by all means at his disposal to help his friends and pupils survive. When experimental work became completely impossible he started reading in other fields around biochemistry, and induced others to extend their knowledge by discussions in small groups. At this time, too, he wrote a short textbook on Physiological Chemistry which was widely used in Holland, running into several editions.

A few months after the liberation of the Netherlands Westenbrink was appointed, as successor to Ringer, to the chair of Physiological Chemistry at Utrecht, which position he had until his untimely death. Westenbrink took over a practically empty laboratory as far as scientists were concerned. Work had not been possible during the last years of the war and Ringer had become ill during this period. Within a few years Westenbrink accomplished what he had set out to do, to fill what he then thought was much space with the noise of biochemists. Some years later the laboratory had become too small.

Westenbrink's artistic temperament and his teaching background flowed together in his personality; on the one hand he was a careful and exact investigator, on the other hand he possessed the imagination necessary for scientific work and especially for inspiring others. His view that art and science are related human activities was well expressed in his inaugural lecture, which he gave the title "Science for Science's sake". In this lecture he showed how modern views had developed and tried to give his view of future development. He made free use of comparisons taken from art and especially visual art. For instance, he compared the knowledge of muscular contraction to the famous uncompleted painting of an Amsterdam canal by Breitner. He possessed the patience necessary for teaching. He only became impatient with sloppy work and sloppy thinking.

Westenbrink had taught biochemistry to medical students in Amsterdam, and had set up a practical course which was later incorporated in the book *Practicum der physiologische Chemie.* In Utrecht, he tried to instil a permanent attachment to research by giving special courses to those students who saw more in biochemistry than a hurdle on their way to becoming a physician. Although, quantitatively, his success was modest, quite a number of Dutch research institutions may be grateful to him for his pupils.

His influence on Dutch biochemistry was immense. In a direct or indirect way he influenced a whole generation of biochemists, to a large part by his personal example. He always worked hard, and never spared himself. He was convinced that time spent on research and on students was never wasted. On the other hand, he gave time and thought to the organization of biochemistry in Holland, and especially to the creation of the best conditions for research. His aim was, as he himself once stated: "I hope that my pupils will be at 36 years, or preferably at 26 years, where I was, comparatively speaking, at 46". When Westenbrink was appointed in Utrecht, biochemistry was only a minor subject within chemistry or biology at Dutch universities. Now, all Dutch universities have programs with biochemistry as major subject. Of course, the whole development of biochemistry has been rapid all over the world, and this has had its repercussions everywhere. Westenbrink, however, possessed the gift of foresight to predict this development and to press – and with success – for the measures to take the new wave of biochemistry in its stride. His gift of foresight may be illustrated by the fact that he told some students in 1945 to learn some genetics, as this might become an important field for biochemists.

He was for a long time Chairman of the Dutch Biochemical Society and during his last year founded the 'Commission of Biochemistry and Biophysics' of the Royal Netherlands Academy of Science. He represented the Netherlands in the Assembly of the International Union of Biochemistry and was a member of Council. He was also very active in establishing international cooperation in biochemistry. His laboratory was often host to speakers from foreign countries and to joint meetings with other societies, especially the Belgian Biochemical Society.

In 1961–62 he was Rector Magnificus of the University of Utrecht. His rectoral lecture, which according to tradition should deal with a topic from the speaker's field and at the same time be intelligible to the general public, was on the place of biochemistry amidst the sciences. It contained a warning to the medical student that the chemists would take the leading positions in medical and biological research and use biologists and physicians as assistants, unless a change in mentality – he even used the words "intellectual structure" – of the medical and biology students took place. Westenbrink was much concerned with the weak position of biochemistry in Holland, compared with other countries. He induced his pupils to work for some time in other countries and quite a number of them have worked outside Holland, on both sides of the Atlantic.

His scientific work in biochemistry started in Amsterdam with an investigation of factors influencing heart muscle. Later he became interested in metabolic diseases. His main work in Amsterdam, however, was on the physiology of thiamin. He studied the influence of thiamin deficiency on the glycogen content of different organs and on tissue respiration. He showed that, in general, tissue respiration was not decreased in thiamin deficiency. It is a reflection on the faith in authority that his article on this problem was not accepted by a German journal, as it was in conflict with a statement by Abderhalden.

His studies also included the so called thiamin-sparing action of fat. After the discovery of cocarboxylase and its identification as thiamin pyrophosphate, his work covered the determination of all forms of the vitamin in different tissues and in yeast. This led to the studies of the yeast carboxylase. The application of yeast carboxylase to the determination of thiamin pyrophosphate in organs and organelles by the Warburg or Cartesian-diver technique led to a number of interesting discoveries. It was shown that the erythrocytes contain much less vitamin per cell than the leucocytes, and lose considerable amounts of it, even after only 5–10 days deficiency. This opened the way to establishing subclinical deficiency states in humans. Moreover, in animals a distinct correlation between thiamin content and degree of functioning of muscles was established. On the other hand, the problem of the so-called vitamin-sparing action of fat was demonstrated to be caused by the heavier loss of vitamin from tissues when carbohydrate metabolism is increased. His aim in the vitamin studies was to establish the biochemical basis of the deficiency, i.e., to describe the deficiency in terms of enzyme malfunctioning. While this has not yet been possible for any vitamin, it led him and his collaborators to studies of the differential loss of thiamin-containing enzymes in deficient animals, especially pigeons, which showed large differences between different enzymes. In all this work, he tried to get at the fundamental processes underlying the symptoms.

He also investigated the absorption of different sugars from the intestine, and showed that, with some sugars this depends on prior intake. This phenomenon is as yet unexplained as is the general phenomenon of specific absorption of glucose, for example.

In his later period in Utrecht, his interest also extended to protein metabolism and to protein structure-function relationships.

This work was devoted to the structure of muscle actin and its conversions in connection with muscular contraction, and to a comparative study of the histones in different organs. In addition, his laboratory became one of the centres of investigation of myeloma proteins, especially Bence-Jones proteins.

Westenbrink's ability to formulate the basic problems clearly and to separate the essential from the non-essential made him an often-consulted adviser on many problems, inside and outside biochemistry. His wisdom was clearly evident in his different functions in scientific organizations. He combined this wisdom with a rare humility. To some extent, in dealing with others, he might be called naïve as he always refused to attribute base motives to others. He completely lacked jealousy and envy and attributed every success to his pupils and the failures to himself. Besides painting, he was interested in literature, especially French and English literature and English history. He himself was a good and fluent writer as his biography of Pekelharing shows, in which he also sketched biochemistry in the nineteenth century in the Netherlands.

His influence on his pupils was great and created bonds of friendship among them with "de prof" or "de baas" as they called him, as centre. He received an honorary doctorate from the University of Aix-Marseille in 1948, the honorary membership of the Dutch Biochemical Society and the membership of the Royal Netherlands Academy of Science.

ELSEVIER PUBLISHING COMPANY

Who was Elsevier, described by Westenbrink as "one of the oldest and greatest of Dutch publishing houses"? In fact at the time when the first issue of BBA appeared, the total staff of the Elsevier Publishing Company amounted to six people, with a yearly turnover in 1946 of 80,000 guilders and in 1947 of 120,000 guilders (at that time about $30,000), equivalent to 720,000 1985 guilders. The director was W.P. Klautz and, in late 1946, Dr. Willem W. Gaade, an organic chemist, joined as scientific worker and Piet Bergmans, a member of a well-known family of booksellers in the south of The Netherlands, as commercial leader.

The history of the Elsevier Publishing Company has been described by P. Vinken [10], the present President of Elsevier-NDU, BV in a foreword to a book *Development of Science Publishing in Europe,* published in 1980 to celebrate the 100th anniversary of the foundation of the Company by the Rotterdam bookseller, Jacobus George Robbers. The House of Elsevier had been founded as a Publishing House in Leiden exactly 300 years previously by Lodewijk Elsevier, a Flemish printer and book-binder from Louvain. In the Seventeenth Century the House of Elsevier had become a true University Press, publishing between two and three thousand titles. In addition to all the major classical works of the early Greeks and Romans, books from some of the most eminent members of Leiden University, including Scaliger, Heinsius (both father and son), Grotius, Snellius and Meursium appeared with the Elsevier family crest – an elm tree next to which stands a hermit.

There is, however, no historical or legal continuity between the House of Elsevier and the Elsevier Publishing Company. The last printing house of Elsevier, in Amsterdam, was closed in 1712. Robbers merely chose its name for the new company and used the Elsevier family crest for its emblem. Nevertheless, the similarity in name has often led to the misconception, also held by Westenbrink, that Elsevier Publishing House is the lineal descendant of the firm established by Lodewijk Elsevier in Leiden in 1580.

Three generations of Robbers – the founder, his three sons and one grandson – were directors between 1880 and 1944*. In 1887, the company moved from Rotterdam to Amsterdam and soon acquired the Dutch-language rights of Jules Verne's *Illustrated Travels* in 57 volumes and of A. Winkler Prins' *Illustrated Encyclo-*

---

*In what follows, I borrow much from the chapter by J.K.W. van Leeuwen, *The Decisive Years for International Science Publishing in the Netherlands after the Second World War* [11].

paedia. Both publications still appear in the Elsevier list. In 1980, the eighth completely revised edition of the Winkler Prins *Encyclopaedia* appeared.

The internationalization of Elsevier started in 1933, when J.P. Klautz became President of Elsevier, a position he held until 1956. Belgium, where the encyclopaedia had an important market, was an obvious first step, but the decisive one for the future of Elsevier was the introduction of books written in English into Elsevier's list. To this end, Klautz acquired the English translation rights to several classical texts and handbooks from the German publishers Thieme Verlag and Akademische Verlagsgesellschaft. Elsevier published the English translation of Paul Karrer's successful textbook, *Lehrbuch der Organischen Chemie,* originally published by Thieme, and of Richter Anschütz's *The Chemistry of Carbon Compounds,* originally published by Akademische Verlagsgesellschaft. An ambitious project, started in 1937, was an encyclopaedia of organic chemistry, following the example of the organic chemist's 'bible' – Beilstein's *Handbuch der Organische Chemie,* published by Springer. Indeed, two former editors of Beilstein, who had left Germany, were editors of the new book. The first volume was not printed until April 1940, but while it was still at the binders Germany invaded Holland and it did not appear until 5 years later.

A London branch of the Company was founded in 1939, and a subsidiary in New York (Elsevier New York) was planned as the combined property of Dekker and Nordmann and Elsevier. Marcel Dekker, one of the owners of Dekker and Nordmann, a bookselling firm in Amsterdam, who was a Ph.D. in biochemistry, moved to New York, where he was joined on the board of Elsevier New York by Dr. Eric Proskauer, previously of Akademische Verlagsgesellschaft. The outbreak of War prevented Elsevier New York functioning as originally intended and Dekker and Proskauer founded the highly successful Interscience Publishers. The owners of Akademische Verlagsgesellschaft managed to escape from Leipzig and set up the also highly successful Academic Press. Both Interscience and Academic Press were to play an important role in biochemical publishing after the War.

During the War, Elsevier ceased to function in Holland. Its German refugee editors were either sent to their deaths in Germany or went into hiding. By the time it got under way again, Klautz had been joined by R.E.M. van den Brink, a young economist, on the board of directors. It was he who, by maintaining a careful cash-flow policy, laid the financial basis for Elsevier's expansion. Gaade and Bergmans were primarily responsible for the remarkable development of scientific publishing by Elsevier immediately after the War. As Bergmans recently (too modestly) put it:

> "The co-operation between the two was ideal, mainly due to the fact that Gaade had no idea of commercial matters and Bergmans even less of scientific."

The *Encyclopaedia of Organic Chemistry* and Karrer's and Richter Anschätz's textbooks served primarily to put Elsevier on the map, but they were not commercially successful. A book programme, however, developed quickly, some running to 15 000 copies, selling mainly in the U.S.A. *Biochimica et Biophysica Acta* was the first journal to be published by Elsevier. How it fared is the subject of the next chapters.

REFERENCES

1 Oppenheimer, C. (1909) *Die Fermente und ihr Wirkungen,* Vogel, Leipzig
2 Oppenheimer, C. (1912) *Grundriss der Biochemie,* Thieme, Leipzig
3 Krebs, H. and Johnson, W.A. (1937) *Enzymologia* 4, 148–156
4 Krebs, H. (1981) *Reminiscences and Reflections,* p. 98, Oxford University Press, Oxford
5 Kalckar, H. (1937) *Enzymologia* 2, 47–52
6 Kalckar, H. (1939) *Enzymologia* 5, 365–371
7 Steyn-Parvé, E.P. (1965) *Biochim. Biophys. Acta* 97, vi–x
8 Cori, C. (1965) *Biochim. Biophys. Acta* 97, xii
9 Gruber, M. (1965) *Biochim. Biophys. Acta* 97, i–v
10 Vinken, P. (1980) in *Development of Scientific Publishing in Europe* (Meadows, A.J., ed.), pp. vii–ix, Elsevier, Amsterdam
11 Van Leeuwen, J.K.W. (1980) in *Development of Scientific Publishing in Europe* (Meadows, A.J., ed.), pp. 251–268, Elsevier, Amsterdam

CHAPTER 2

The first volume

The first issue of BBA appeared on 30 January, 1947, Westenbrink's 46th birthday, as P. Bergmans recalls, a real day of celebration for Elseviers' staff of six. The Editorial Board comprised W.T. Astbury (United Kingdom), A.E. Braunstein (U.S.S.R.), C.F. Cori (U.S.A.), Cl. Fromageot (France), K.U. Linderstrøm-Lang (Denmark), H.G.K. Westenbrink (The Netherlands) and R.W.G. Wyckoff (U.S.A.). There was also an Advisory Board, composed of J.D. Bernal (United Kingdom), J. Brachet (Belgium), T. Casperson (Sweden), C.R. Harington (United Kingdom), A.J. Kluyver (The Netherlands), H.A. Krebs (United Kingdom), A. von* Muralt (Switzerland), A.J. Oparin (U.S.S.R.), J. Roche (France), M. Sreenivasaya (India), D.L. Talmud (U.S.S.R.) and H. Wu (China).

In the 1948 catalogue of Elsevier, *Biochimica et Biophysica Acta* was described as 'International Journal of Biochemistry and Biophysics, Revue Internationale de Biochimie et Biophysique, Internationale Zeitschrift für Biochemie und Biophysik.'

---

*Incorrectly printed as De.

"The revival of normal scientific activity in Europe has emphasized the lack of appropriate media of publication in the field of biochemistry and biophysics. For this reason, Elsevier Publishing Co., Inc., in conjunction with an international group of outstanding scientists, has inaugurated an international journal devoted to this field of research. Interscience Publishers, Inc., New York, are sponsoring the journal for the United States.

Both the Editorial Board and publishers consider the new journal a symbol of the re-establishment of international cooperation in the post-war period. American and British scientists will no doubt like to utilize this new journal as a means of presenting their new ideas and results to their European colleagues. *Biochimica et Biophysica Acta* is destined to bridge national gaps and divergences, and aims at a true international exchange of scientific information in the field of biochemistry and biophysics.

Special emphasis is being given to the inclusion of biophysicists on the Editorial Board. Until now, biophysical papers have been widely scattered. The new journal makes a special effort to attract them, guided by the principle that the biophysicist attempts to achieve the same result as the biochemist, but by different methods.

The journal publishes original papers, review articles, surveying progress in particular fields, brief letters to the editors in English, French and German, with a brief summary of each article in these three languages. It appears bi-monthly, and averages about 100 pages per issue."

The subscription was $9.00, post free.

In the Westenbrink Memorial Number, R.W.G. Wyckoff [1] described Westenbrink's rôle and the editorial policy of the journal in the following words:

"I am not sure how many of the readers and contributors to the Acta, and even of the newest editors, realize how fully Westenbrink was responsible for their journal. At the outset he managed it singlehanded while steadily refusing the title of managing editor. It was he who arranged matters with Elsevier and dealt directly with those who submitted manuscripts. All we other editors did was to accept papers and give him our best advice as to their suitability. At the time the Acta was being started the resurgence of scientific work following the war was already beginning to yield its inevitable spate of new papers; and the burden they put on existing journals was giving much support to those who favored the dehumanizing of scientific publications – the creation of boards of editors who would rework submitted manuscripts into a common mold, accepting those they decided were correct and seeing that were rewritten till the individuality of the man who had done the research was lost in this common pattern of reporting. The Acta was founded on an editorial policy the antithesis of this. Westenbrink believed that a man's manuscript was as much his own as the work described and merited the same respect. An editor was needed to decide if the work lay within the province of the journal, if it made sense and if it appeared to be a genuine contribution to knowledge; he also had the obligation to see that the paper was sufficiently well written to be read with understanding, to point out glaring errors that might have escaped the author and to ensure that the paper did not occupy more space than was required to convey its results. Beyond this an editor should not go. Scientists must be treated as adults with an adult responsibility towards their work. If that was lacking, if a man made more than the unavoidable minimum of errors in his research and was careless in its description, his colleagues were the ones to bring him to task. They were his judges and not the editors or anonymous referees deciding whether or not a reasonable-appearing piece of work should be published. Credit for the Acta's present obvious success should go in a very large measure to the wise and human way in which Westenbrink maintained this editorial policy."

Westenbrink's successors have tried to follow this editorial philosophy, although it became increasingly difficult with the great increase in the number of papers handled, involving a large number of people at different stages, to avoid some degree of dehumanization in the relations between authors and the journal.

It is difficult now to reconstruct from the bound Volume 1 of *Biochimica et Biophysica Acta* found on the library shelves precisely what comprised the first of the six issues belonging to this volume. The honour of a reference to Volume 1, p. 1, belongs to Stig Veibel and Gregers Østrup from the Chemical Laboratory of the University of Copenhagen for their paper entitled *Investigation on the β-galactosidase of alfafa seed emulsin* [2], received on March 26, 1946. The second paper, received on the same date, is by J. Brachet of the

University of Brussels, then a member of the Advisory Board, later (and still) a member of the Editorial Board, together with R. Jeener on *Protéines de structure de Szent-Györgyi et thymonucléohistone* [3].

The third paper is by Theodor Bücher and Jacob Kaspers from the Kaiser Wilhelm Institut für Zellphysiologie, then in Liebenberg, Germany, where Otto Warburg's laboratory had been evacuated from Berlin. This paper is on *Photochemische Spaltung des Kohlenoxymyoglobins durch ultraviolette Strahlung (Wirksamkeit der durch die Proteinkomponente des Pigments absorbierten Quanten)* [4]. In his reminiscences, Krebs [5] relates the circumstances by which it came to be published in BBA. According to Krebs, Bücher's paper, as well as three others that were published in a later number of the first volume of BBA, had been accepted for publication by the *Biochemische Zeitschrift* and had reached the proof stage, but the proofs were lost when the printing house in Leipzig was bombed.

> "As the Russian army advanced on Liebenberg, Bücher moved his family westward to the British zone of occupation and settled in a village not far from Lübeck. He got work in a chemical factory. One day, members of the British Intelligence Service happened to see him dismantling some electronic equipment and suspected him of trying to set up radio communications. Bücher was taken for interrogation but succeeded in convincing his questioners that he was not a secret agent but a research scientist who, moreover, was anxious to have their help in getting three of his unpublished papers to Britain. He asked if they could be sent to Professor Donnan, care of the Royal Society. Through Donnan, these papers reached me at Sheffield and once again Canon Austen [the husband of Krebs' secretary; he was then serving as an army chaplain with the British Forces in Germany – E.C.S.] was instrumental in getting Bücher and myself in touch. I submitted the papers for publication to the *Biochemical Journal* and received a letter, dated 22 January 1946, from the Chairman of the Editorial Board, stating 'The Committee of the Biochemical Society have directed the Editorial Board that, until further notice, the Editors should not accept papers submitted from ex-enemy countries. In the circumstances it will obviously be better for Dr. Bücher to publish in a neutral country or to wait until publication begins again in Germany.'

In the event they were published not in a neutral country, but in Holland. This came about because I had an opportunity to submit the papers to a new journal, *Biochimica et Biophysica Acta*, and in July 1946 they were accepted for publication through the courtesy of the Editor, Professor Westenbrink and of the publishers, the Elsevier Company. The papers appeared in the first volume of *Biochimica et Biophysica Acta* and were widely recognized as being of great importance."
[Reproduced from [5] with permission.]

Indeed, these four papers were in the nature of a scoop for BBA. One of them, *Über ein phosphatübertragendes Gärungsferment* [6], in particular, describing the isolation, crystallization and properties of what is now known as 3-phosphoglycerate kinase (in the 1950's known as 'Bücher's enzyme') is a classical paper in glycolysis and phosphorylation. The other two were on the molecular weight of enolase [7] with a companion paper on the validity of applying the Rayleigh equation for the measurement of the molecular weight of proteins by light scattering [8].

Other well-known authors in the first volume include P. Desnuelle [9, 10], J. Roche and R. Michel [11–13], F.L. Breusch [14], L. Massart [15, 16], H.G.K. Westenbrink and E.P. Steyn Parvé [17, 18], H. Holter and K. Linderstrøm-Lang on the micro-Kjeldahl method [19], R.W.G. Wyckoff [20], J.M. Wiame [21], J. Courtois [22, 23], A.E. Braunstein [24], H. Veldstra [25], W.T. Astbury [26], A. Engström [27], H. Chantrenne [28], C. Fromageot [29], F. Haurowitz [30], K. Bailey [31], S.V. Perry [31, 32], and W. van Iterson [33, 34].

One paper by J. Groen, W.A. van den Broek and H. Veldman [35] reminds one that the tragedy of the War was still very recent. A footnote records that both the second and third authors died in 1945: van den Broek was shot down in the street by a Nazi agent; Veldman died in a German concentration camp.

A feature of the first volume was the beautifully reproduced micrographs of tobacco mosaic virus [20], actin [32], chloroplasts [34] and bacteria [33].

The volume concluded with an appeal by UNESCO to help the

International Bureau of Physico-Chemical Specimens at Brussels to collect pure chemical compounds.

The first volume firmly established the international basis of BBA. Ten countries were represented. Not surprisingly perhaps for the first volume, more papers (34%) came from the Netherlands than from other countries. They were followed by France, 17%; Belgium, 11%; U.S.A., 9%; Germany and U.K., each 8%; Denmark, 6%; Turkey, 4% and U.S.S.R. and Sweden, each 2%. Most of the papers (60%) were in English, but the numbers in French (27%) and German (13%) were much higher than in subsequent years.

REFERENCES

1 Wyckoff, R.W.G. (1965) *Biochim. Biophys. Acta* 97, xi–xii
2 Veibel, S. and Østrup, G. (1947) *Biochim. Biophys. Acta* 1, 1–12
3 Brachet, J. and Jeener, R. (1947) *Biochim. Biophys. Acta* 1, 13–20
4 Bücher, T. and Kaspers, J. (1947) *Biochim. Biophys. Acta* 1, 21–34
5 Krebs, H. (1981) *Reminiscences and Reflections*, pp. 149–150, Clarendon Press, Oxford
6 Bücher, T. (1947) *Biochim. Biophys. Acta* 1, 292–314
7 Bücher, T. (1947) *Biochim. Biophys. Acta* 1, 467–476
8 Bücher, T. (1947) *Biochim. Biophys. Acta* 1, 477–483
9 Desnuelle, P. and Antonin, S. (1947) *Biochim. Biophys. Acta* 1, 50–60
10 Desnuelle, P. and Rovery, M. (1947) *Biochim. Biophys. Acta* 1, 497–505
11 Roche, J. and Michel, R. (1947) *Biochim. Biophys. Acta* 1, 335–356
12 Roche, J., Michel, R. and Lafon, M. (1947) *Biochim. Biophys. Acta* 1, 453–466
13 Nguyen-Van Thoai, J., Roche, J. and Roger, M. (1947) *Biochim. Biophys. Acta* 1, 61–76
14 Breusch, F.L. and Tulus, R. (1947) *Biochim. Biophys. Acta* 1, 77–82
15 Massart, L. and Hoste, J. (1947) *Biochim. Biophys. Acta* 1, 83–86
16 Deley, J., Peeters, G. and Massart, L. (1947) *Biochim. Biophys. Acta* 1, 393–397
17 Westenbrink, H.G.K. and Steyn-Parvé, E.P. (1947) *Biochim. Biophys. Acta* 1, 87–94
18 Westenbrink, H.G.K., Steyn-Parvé, E.P. and Veldman, H. (1947) *Biochim. Biophys. Acta* 1, 154–174

19 Brüel, D., Holter, H., Linderstrøm-Lang, K. and Rozits, K. (1947) *Biochim. Biophys. Acta* 1, 101–125
20 Wyckoff, R.W.G. (1947) *Biochim. Biophys. Acta* 1, 139–146
21 Wiame, J.M. (1947) *Biochim. Biophys. Acta* 1, 234–255
22 Fleury, P. and Courtois, J. (1947) *Biochim. Biophys. Acta* 1, 256–269
23 Courtois, J. (1947) *Biochim. Biophys. Acta* 1, 270–277
24 Braunstein, A.E., Nemchinskaya, V.L. and Vilenkina, G.J. (1947) *Biochim. Biophys. Acta* 1, 281–291
25 Veldstra, H. (1947) *Biochim. Biophys. Acta* 1, 364–378
26 Astbury, W.T. and Spark, L.C. (1947) *Biochim. Biophys. Acta* 1, 388–392
27 Engström, A. (1947) *Biochim. Biophys. Acta* 1, 428–433
28 Chantrenne, H. (1947) *Biochim. Biophys. Acta* 1, 437–448
29 Fromageot, C. and Clauser, H. (1947) *Biochim. Biophys. Acta* 1, 449–451
30 Haurowitz, F. and Tekman, S. (1947) *Biochim. Biophys. Acta* 1, 484–486
31 Bailey, K. and Perry, S.V. (1947) *Biochim. Biophys. Acta* 1, 506–516
32 Perry, S.V. and Reed, R. (1947) *Biochim. Biophys. Acta* 1, 379–388
33 Van Iterson, W. (1947) *Biochim. Biophys. Acta* 1, 527–548
34 Algera, L., Beyer, J.J., Van Iterson, W., Karstens, W.K.H. and Thung, T.H. (1947) *Biochim. Biophys. Acta* 1, 517–526
35 Groen, J., Van den Broek, W.A. and Veldman, H. (1947) *Biochim. Biophys. Acta* 1, 315–326

CHAPTER 3

The early years

FINANCIAL PROBLEMS

In its early years, BBA was a great source of financial worry for the Publishers. In 1947 it contributed only f 15,700 to the turnover of the Company (of a total of f 120,000) and made a loss of not less than f 34,600 (see Table I). By 1950, the turnover had more than quadrupled, but there was still a loss of f 10,200 and the accumulated loss had reached f 102,500, a very large sum for those days, equivalent to more than 500,000 1985 guilders. It proved extremely difficult to sell an unknown journal from a publisher unknown in this field, even though distribution in the United States was entrusted to Interscience Press. The editors also had great difficulties, despite the promising start, in obtaining sufficient copy for the second volume. This is hardly surprising. It was less than two years since the end of the War, and although biochemists in many countries were back in their laboratories, it took time before the work was ready to be published. Nevertheless, a second volume appeared in 1948 and by 1949 sufficient papers were being received to warrant the publication of two volumes.

Table I

Turnover and loss of BBA in the first four years

Year	Turnover (Dfl)	Loss (Dfl)
1947	15,700	34,600
1948	18,400	31,000
1949	23,200	26,700
1950	71,300	10,200
		Total 102,500

During this period, Klautz made two attempts to sell the financial burden that BBA had become to the company, first to Interscience Press, who, however, since they were the distributors in U.S.A., were aware of its poor prospects. Cleaver-Hume Press in London also declined.

Fortunately, it was clear by 1950 that at least the rate of loss had lessened, but it was not until 1954 that the accumulated loss was cleared and Elsevier made its first profit. In fact, this was confirmation of an old saying in the publishing world that 7 years are required before the first profit is made.

The relatively slow growth of BBA between 1947 and 1955 is illustrated in Fig. 2. The two volumes issued in 1949 could not be maintained, the next three volumes being spread over 1950 and 1951. In 1951 an important decision was taken – namely, to accept Short Communications and Preliminary Notes. As will be described in the next section, this step was decisive for the future growth of BBA. In 1952, two thick volumes appeared and in the period 1953 to 1955, the journal had become established, with three volumes per year.

The number of editors was gradually increased. Already on the second volume, Brachet had been moved from the Advisory Board to the Editorial Board. In 1951, the first editor from Germany –

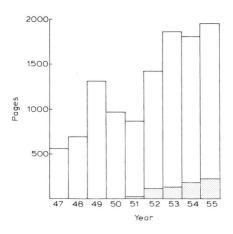

Figure 2. The size of BBA, 1947–1955. Shaded area represents Preliminary Notes and Short Communications.

H.H. Weber of Tübingen, later Heidelberg – joined the Editorial Board, to be followed in 1955 by E. Chargaff (New York) and R. Wurmser (Paris).

Two 'festschrift' numbers appeared, one in 1953 (Vol. 12) in honour of Otto Warburg's 70th birthday and one in 1956 (Vol. 20) in honour of Carl and Gerty Cori's 60th birthday. No festschrift numbers have been published since 1956, since they were found to cause too much disruption of the usual production of the journal.

SHORT COMMUNICATIONS AND PRELIMINARY NOTES

In Volume 7 (1951) Short Communications and Preliminary Notes appeared in BBA, at first grouped together, later (from Volume 20) under separate headings. Whether this was Westenbrink's idea or that of one of the other editors or the publisher I do not know. Also I have been unable to trace how these categories of paper were first described. In the first 'Notice to Contributors', which did not appear until 1959, three categories were distinguished:

> *Normal-length papers.* Although considerable freedom regarding the form and lay-out of normal-length papers is allowed, authors should consult a current issue in order to make themselves familiar with the typographical conventions, lay-out of tables, citation of references, etc. A summary should be included. Papers may be published in English (either American or British spelling is permitted), French or German. An additional English summary will be prepared by the publisher, when the paper is in French or German.
>
> *Short Communications* are concise but complete descriptions of a small rounded-off investigation, which it is not intended to include in a later paper. Suitable topics are, for example, an analytical or preparative method, or the report of work which is on a side path from the main line of investigation. A paper will not be accepted as a Short Communication if it is considered that it can be more satisfactorily incorporated in a longer paper, when the investigation has progressed further. Short Communications will be printed in the same type as used for normal-length papers, but no sub-division into sections is in general necessary. No summary is required. A Short Communication should not occupy more than three pages of the journal. The publication time will be about the same as that of normal-length papers.
>
> *Preliminary Notes* are brief reports of work which has progressed to the stage when it is considered that the results should be made known as soon as possible to other workers in the field. It would help the Editors if the author, in a covering letter, gave his reasons for believing that publication in this form is urgent. The material can be later incorporated in a normal-length paper in this or in another journal. The size of print and lay-out of Preliminary Notes will be the same as that of Short Communications. The maximum length allowed will be two printed pages. If no correspondence with the author is required, it is expected that Preliminary Notes will be published about two to three months after receipt.
>
> Normal-length papers and Short Communications should be presented so that they are intelligible to the ordinary reader of *Biochimica*

et Biophysica Acta, and should include sufficient information to allow the experiments to be repeated. Preliminary Notes should contain sufficient information so that workers in the same field will understand the sort of experimental evidence on which a conclusion is based.

In the above announcement, it is stated that both Short Communications and Preliminary Notes will be printed in the same type as used for normal-length papers. This was, in fact, a recent development announced by the Publisher at a meeting with some members of the Editorial Board during the Fourth International Congress of Biochemistry in Vienna in 1958. Until then, for financial reasons, these categories had been printed in small type.

The introduction of Preliminary Notes and Short Communications was a big success. The great post-war expansion of biochemistry was under way in the 1950's, competition was becoming fierce and many biochemists wished to make their discoveries known before they could appear in normal-length papers. Letters to *Nature* had long been used for this purpose, but this journal had to cater for the whole field of natural science and although what are now recognized as historical papers in biochemistry and molecular biology were published in *Nature* at that time, it could not cope with all the demands of the biochemist. *Science* has never been as important as *Nature* in this respect. The *Journal of Biological Chemistry* published a small number of letters and the Proceedings of meetings of the Biochemical Society, published as an Addendum to the *Biochemical Journal*, served somewhat the same purpose. There were, however, at that time no biochemical journals specialized in the rapid publication of short notes.

The market was, therefore, wide open for the BBA Preliminary Notes and Short Communications. Many important announcements first appeared in these papers. To give just a few examples from one field – mitochondrial research – the role of coenzyme Q or ubiquinone [1], the identification of diaphorase with lipoamide dehydrogenase [2], the identification of firmly bound copper in cytochrome *c* oxidase [3], the utilization of 'high-energy intermediates' of oxidative phosphorylation [4], the first soluble 'factor' of oxidative phosphorylation [5], protein synthesis in mitochondria [6] even when bacterial contamination is rigorously excluded [7], the identification of a mitochondrial DNA with unique renaturation behaviour [8] consisting of closed circles [9] and the use of antibiotics such as chloramphenical and erythromycin to inhibit specifically synthesis of mitochondrial components coded for by the mitochondrial genome [10] were all reported first in Preliminary Notes in BBA. By 1955, 17% of the total number of pages were devoted to Short Communications and Preliminary Notes. Since the average length was only 1–2 pages, the number of articles listed in this category in the Table of Contents soon approached the number of regular papers.

There is little doubt that the introduction of Preliminary Notes and Short Communications contributed to the large increase in the number of readers of BBA. The number of subscribers increased from 630 in 1950 to 920 in 1952. The success of the shorter papers stimulated many leading authors to send in papers not only in this category but also regular papers. Thus, it had a beneficial effect on the general standard of the journal.

However, the success of the Prelininary Notes brought problems. First, the large increase in the number of papers that, by their very nature, had to be dealt with rapidly meant that it was no longer possible for Westenbrink to carry the whole load, especially as his standing in the academic world in the Netherlands meant that many other tasks fell upon him, including that of Rector of the University of Utrecht. Secondly, the success of the Short Communications and Preliminary Notes was undoubtedly a stimulus to other journals to introduce a similar section and to a rival publishing company (Academic Press) in 1960 to launch a new journal devoted only to this type of publication – *Biochemical and Biophysical Research Communications*. In itself, this was welcome for BBA, since it relieved the pressure of papers submitted in this category. In 1960, BBA received 626 Regular Papers, 252 Short Communications and 316 Preliminary Notes. However, since rapid publication, particularly of Preliminary Notes, was one of their most attractive features, it became increasingly difficult to compete with the new journal for the most important Preliminary Notes from America, where biochemical research was concentrated, particularly in the 1960's. The decline in the efficiency and increase in the costs of the postal services that developed at this time were also a great handicap for a journal that resolved to retain a refereeing system and, if possible, offer authors the possibility of correcting proofs.

Biochemical and Biophysical Research Communications was one of the first to introduce the publication of typescripts made camera-ready by the authors, which obviates the need for proofs. There were several reasons why BBA did not follow suit. In the first place, our authors came from all over the world, including countries where typewriters suitable for camera-ready copies were not always available and where the knowledge of one of the three languages in which BBA was published was not such that papers could be accepted without language correction. Secondly, we felt that it would be deleterious to the appearance of BBA, whose prime role was to publish regular papers, to tag onto it a section in typewriter script. Thirdly, Slater, who took over the editing of Preliminary Notes and Short Communications in 1956 (see the next chapter) found it a frustrating prospect not to be able to make editorial 'improvements', even in questions of nomenclature.

Unfortunately, the inherent difficulties in competing with *Biochemical and Biophysical Research Communications* were compounded by problems that Elsevier was having with rapid publication. Despite the extra delays entailed in sending manuscripts to Melbourne and back to Amsterdam during Slater's stay there in the middle of 1963 (see p. 45), he found on his return that by far the main contribution to the long publication time was between the date on which the first proof was printed and the date of publication. After pointing out some glaring cases of publication delays, he concluded (translation from Dutch):

> "I am aware of the strong 'over-employment' at present in the Netherlands. However, I think that I must warn you that the present long publication time for Preliminary Notes will have the undesirable consequence that more of the better American Preliminary Notes will appear in *Biochemical and Biophysical Research Communications*."

Indeed, that the rapid increase in salaries and the over-extended labour market were causing problems for the Publisher was made very clear by a long letter from Bergmans to the Managing Editors in the beginning of 1965, explaining why the 'lay-out' of Tables had been drastically altered. Bergmans pointed out that the wage explosion that started in 1963 had hit the labour-intensive printing industry in Holland particularly hard and had brought the realization that the industry as a whole was technically backward and that rationalization (including standardization) was drastically necessary. To follow customary procedures of set typing would require an immediate increase of 30–40% in printing costs, which was not acceptable to the Ministry of Economic Affairs, who instead told the printers to industrialize their profession, and this required in the first instance a standardization of typographical procedures. Bergmans' letter had the important effect of bringing home to the Managing Editors that there was more to bringing out a journal than just editing it. It was a good example of the advantage of keeping the two partners – Editors and Publishers – informed of each other's difficulties.

One of the reasons for the delay in publication time was the sectionalization of BBA, which meant that some sections were appearing at quite long intervals. If a Preliminary Note belonging to one of these sections just 'missed' an issue, it could be up to 2 months before another number of this section appeared.

The number of Preliminary Notes received peaked around about 1960, but it was not until the end of 1967 that the question of discontinuing this category was considered, as a result of discussions held with editors during the Seventh International Congress in Tokyo in the previous summer. Shortly afterwards (1968) *FEBS*

Letters started to compete for short papers. By May, 1969, Preliminary Notes accounted for only 5% of the output, and during the last 2 months 59-70% of Preliminary Notes submitted had been refused. Since there were relatively few articles warranting urgent publication, a single short-paper category would be adequate.

For the November, 1969 meeting of Managing Editors, a Memorandum was prepared by J. Geelen summarizing comments made by Editors at various meetings and by letter in response to a passage in the Annual Report to the Editorial Board for 1967. Many of the newer editors seemed to be unaware of the original distinction between the two categories of short papers. The distinction had in fact become blurred by the appearance in many other journals of 'short notes', 'preliminary communications', etc. At this meeting, it was decided to stop both categories and to introduce a new one, that later, after further consultation with the Editorial Board, was given the name of BBA Reports. The last Short Communications and Preliminary Notes appeared in Volume 224 (1970) and the first BBA Reports in Volume 225 (1971). Interestingly, the number of subscribers to BBA reached its peak in 1970.

BBA Reports were envisaged to combine the advantages of the two earlier categories without their disadvantages. Rapid publication was promised, but the report was not preliminary, so that the data could not be republished in a full paper. Indeed, the idea of such double publication was now frowned on in biochemical circles*. It was hoped optimistically that the rapid publication promised would attract papers reporting 'exciting' new developments of the type that were appearing in the *Proceedings of the National Academy of the U.S.A.* Not surprisingly perhaps, this hope has not been realized and, as a result of a decision taken in the March, 1981

---

*As Slater put it in a circular to members of the Editorial Board on January 27, 1970: ". . there has been a change in the conventions of scientific publication. Whereas at one time, it was considered normal and even desirable to repeat in the full paper data already published in a preliminary form, many readers consider this now as double publication and, therefore, undesirable." The undesirability of double publication is now generally accepted by reader and Publisher alike.

Meeting, the description of BBA Reports in the 1982 issue of *Information for Contributors* is quite similar to that originally given for Short Communications. Specifically it is stated that

> "A BBA report, although brief, should be a complete and final publication, and figures and tables from the Report should not be included in a later paper."

It is no longer logical to offer rapid publication of BBA Reports, and indeed it is not necessary now that the publication of Regular Papers has been brought back to about 5 months.

However, memories are long-living and many authors and even editors still confuse BBA Reports with the old Preliminary Notes. With hindsight, some of the decisions taken at the end of the 1960's do not now appear to have been wise. Some confusion might have been avoided if Short Communications had been retained, instead of introducing BBA Reports.

This section on Short Communications and Preliminary Notes has now overrun the period covered by this chapter – 'The Early Years'. In the following chapter, we shall return to the middle 1950's, when BBA was already established.

REFERENCES

1 Crane, F.L., Hatefi, Y., Lester, R.L. and Widmer, C. (1957) *Biochim. Biophys. Acta* 25, 220–221
2 Massey, V. (1957) *Biochim. Biophys. Acta* 30, 205–206
3 Takemori, S., Sekuzu, I. and Okunuki, K. (1960) *Biochim. Biophys. Acta* 38, 158–160
4 Snoswell, A.M. (1961) *Biochim. Biophys. Acta* 52, 216–218
5 Linnane, A.W. (1958) *Biochim. Biophys. Acta* 30, 221–222
6 Simpson, M.V. and McLean, J.R. (1955) *Biochim. Biophys. Acta* 18, 573–575
7 Kroon, A.M., Saccone, C. and Botman, M.J. (1967) *Biochim. Biophys. Acta* 142, 552–554
8 Borst, P. and Ruttenberg, G.J.C.M. (1966) *Biochim. Biophys. Acta* 114, 645–647

9 Van Bruggen, E.F.J., Borst, P., Ruttenberg, G.J.C.M., Gruber, M. and Kroon, A.M. (1966) *Biochim. Biophys. Acta* 119, 437–439
10 Huang, M., Biggs, D.R., Clark-Walker, G.D. and Linnane, A.W. (1966) *Biochim. Biophys. Acta* 114, 434–436

CHAPTER 4

The second decade

SLATER JOINS WESTENBRINK

For practically the entire first decade of BBA's existence, its Managing Editor (in function but not in name) was Westenbrink. He received papers mostly directly, sometimes from one of the members of the Editorial Board, and, with the help of a part-time secretary (Mrs. Lida Koning), sent the papers off for review. He also carried out all correspondence with authors and would-be authors, and forwarded acceptable papers to the Publisher, who looked after sub-editing, printing and distribution.

The expansion of BBA, particularly of the section dealing with Short Communications and Preliminary Notes, made some assistance necessary. Just before Slater departed for his summer vacation in 1956, after completing his first full academic year in the University of Amsterdam, he was surprised by an urgent request on the telephone from Gaade for a meeting to discuss a matter of importance. At this meeting, Gaade, also speaking on behalf of Westenbrink, asked Slater if he would take over the task of Managing Editor for Short Communications and Preliminary Notes. Slater was overwhelmed by this request, and did not give a direct answer. During his vacation in Cambridge, he did what he had always done since 1946 when he was undecided about an important scientific or academic matter – he asked his teacher, Professor David Keilin, for advice, fully expecting that this would be to forget about editorial duties and concentrate on building up a biochemical department in Amsterdam. To Slater's surprise, Keilin advised acceptance of the offer. As chief editor of the journal *Parasitology*, Keilin considered that an editorial function was an intellectual stimulus and would give scientific contacts of value to the editor's scientific development.

Further consultation with Gaade, Bergmans and Westenbrink resulted in an appointment per letter of October 4, 1956, from Bergmans, to "handle the section of short communications and preliminary notes." In addition to the question of an honorarium and reimbursement of postal expenses

> "we shall see in how far Elsevier will have to assist you in the sector of secretarial help. This will be a matter of further consideration after your having say 3 or 6 months experience in the editorial work."

During 1957, Westenbrink and Slater ran the different sections practically independently of one another, dealing directly with the authors and sending on acceptable manuscripts and corrected proofs to Gaade for publication. Slater, in particular, spent a lot of time correcting proofs. At the end of 1957, a visit by Professor Cl. Fromageot, one of the founder editors, to the Netherlands provided a good occasion for a discussion of policy with the Publisher. The meeting, which was held in Utrecht on December 3, 1957, was attended by Bergmans, Gaade and Mr. B.H. Mulder for the Publisher and Westenbrink, Slater and Fromageot. It was the forerunner of meetings of this nature which are still held regularly (now usually twice a year).

Mulder, who was now responsible under Gaade for BBA from the Publisher's side, prepared a short report of this meeting. The topics covered read like an agenda for a meeting in 1985, dealing with Instructions to Authors, delay in publication, redrawing of figures, size of BBA, additions to and replacements of members of the Editorial and Advisory Boards. An important minute reads

> "Careful scrutiny of the manuscripts, correspondence in connection with them, calculations of their length, etc., makes increasing demands on Prof. Slater's and Prof. Westenbrink's time. Mr. Bergmans is prepared to consider engagement of secretarial help at Elsevier's expense."

It was also decided to add to the Advisory Board a Japanese representative, and Slater was asked to contact Egami (Nagoya), who duly joined the Board.

Indeed it was at about this time that the rise in Japanese biochemistry was making itself felt in the papers being submitted to BBA. The increase in biochemical competence was at that time not yet matched by the same competence in the English language and this gave difficulties for language correctors employed by Elsevier. In one case, Slater completely rewrote a paper for a Japanese author, who in consequence became a life-long friend.

Slater was now a member (later Secretary) of the Biochemical Nomenclature Committee of the International Union of Pure and Applied Chemistry and was spending a lot of time fighting the increasing use of unnecessary abbreviations and the sloppy use of symbols in the biochemical literature (for example, μM for micromoles instead of micromolar). His corrections caused problems for the sub-editing staff in Elsevier and a number of letters passed between Gaade and Slater, culminating in a letter dated January 27, 1958 from Slater to Gaade which "in order to avoid future misunderstanding", set out the agreed division of responsibilities between himself and Gaade's editorial staff concerning editing of Short Communications and Preliminary Notes. It is strange now to read that Slater was still correcting all proofs (this was done at odd hours, for example, during dull moments in faculty meetings). This part of the letter reads –

> "I shall check proofs received from authors or, when necessary in order to avoid publication delays, before they are received from authors. It is to be understood, however, that my proof reading must be incomplete, since I am not in the possession of the manuscript. Therefore, the proofs must also be carefully read by your staff. I undertake not to make more alterations on the proof than the few you would consider acceptable from an author, except when additional information has been received from the author after I have sent you the manuscript and I consider that it is desirable to make the change.
>
> At the same time, it should be understood that I cannot guarantee never to make a mistake when editing a manuscript. It might be necessary to correct such a mistake on the proof. Since I do not have the manuscript, it is often not possible to know whether it is I or the printer who has made the mistake. I shall assume that mistakes are never made by your staff."

The Fourth International Congress of Biochemistry in Vienna provided the opportunity for Westenbrink and Slater together with representatives of Elsevier to meet with members of the Editorial Board. Braunstein, Chargaff and Cori were able to accept the invitation to a luncheon. This started a tradition that has been continued at most international congresses and some FEBS Meetings. Meyer, who hosted the meeting for Elsevier, sent a report to all members of the Board.

An important decision was taken at this meeting. Up till then authors were able to submit papers to any member of the Editorial Board. Some editors complained that this imposed a heavy burden on them, and moreover they received the blame when papers submitted through them were refused. It was decided, therefore, to ask authors to submit all papers to a 'neutral' address, and that, in view of the large number of papers being received from U.S.A., one of the 'neutral' addresses should be in that country. In subsequent discussions in Amsterdam, however, the practical difficulties inherent in the long distance between Publisher and Editorial Centre became fully realized and it was decided to designate Westenbrink, Laboratory of Physiological Chemistry, Vondellaan 24a, Utrecht for normal-length papers and Slater, Laboratory of Physiological Chemistry, J.D. Meijerplein 3, Amsterdam for Short Communications and Preliminary Notes. It is only in recent years, with the development of modern communication methods, that it has become feasible to place Editorial Offices in other countries.

A Notice to Contributors, explaining the differences in the three categories of papers and instructions on preparation and submission of manuscripts was prepared by Slater for printing on the inside cover of the journal. This Notice to Contributors was first published in Vol. 30 No. 2 (1958), and manuscripts started to stream into Vondellaan and J.D. Meijerplein soon afterwards. In the European summer of 1959, however, Slater had congresses to attend in Australia and Japan, with a fairly long gap in between which he spent visiting his parents in Australia. Consequently, authors and would-be authors of Short Communications and Preliminary Notes were surprised by notes sent from Arcadia, Magnetic Island, North Queensland, Australia, just as the post office there was overwhelmed with large envelopes completely covered with Dutch postage stamps. Slater also carried on with BBA when he visited Melbourne for several months in 1963. A well-known Melbourne biochemist was thrown into a panic by a telephone call which he thought was from Amsterdam, but in fact was from the other side of the campus at Melbourne University.

Also attending the Vienna meeting was Dr. J.F. Remarque, who had just taken up an appointment with Elsevier after having obtained his Ph.D. in biochemistry with Westenbrink. He was destined to play an important role during the expansion period of BBA. In November 1960, he announced Elsevier's plans to open up the possibility to subscribe to certain single issues of BBA containing only papers dealing with a specialized field. The two fields chosen were 'nucleic acids and related topics' and 'enzymology'. This was the beginning of the sectionalization of BBA. Slater was not at all happy with this idea, but now concedes that this has been very successful.

Another innovation by Remarque, which, however, was less successful, was the introduction in 1961 of BBA Previews, consisting of the summaries of papers and distributed several months before the journal itself. Another excellent idea, which also unfortunately did not take on, was to offer reprints for sale to readers, thereby putting the burden of the cost where it belongs – on the reader, rather than the author.

Slater satirized the admirable but sometimes in his view a little too aggressive promotion activities of the publisher in the following mock advertisement:

> "A NEW ELSEVIER SERVICE TO BBA READERS – INSTANT BBA ARTICLES"
> Do you have trouble with digesting BBA articles?
> The scientific department of Elsevier SCIENTIFIC Publishing Company, after years of research, has found the solution – INSTANT BBA ARTICLES. This is a powder which instantly dissolves in 40% ethanol (w/v) to give a crystal-clear instantly digestible solution – green for articles with an enzymological content, yellow for fatty articles, greenish-yellow for articles dealing with the enzymology of lipids, yellowish-green for articles dealing with lipid components of enzymes."

Slater's frustration at the difficulty of allocating articles to their proper section is also revealed here.

THE 'THIRD' MAN

By now BBA was growing very rapidly and it was necessary greatly to increase the size of the Editorial and Advisory Boards. Among those joining the latter at the beginning of 1962 were two young Dutch biochemists – L.L.M. van Deenen from Utrecht and M. Gruber from Groningen.

The first of the regular meetings between Publisher and the

Managing Editors took place on May 9, 1963 in Westenbrink's office in Utrecht. It was attended by Westenbrink, Slater, Remarque, J. Geelen and J.G.M. Kleijn. The latter was now 'business manager' of BBA, being responsible for the printing and distribution.

Already in this first meeting, Slater expressed his concern with the growth of BBA, not only of the number of articles (900 to be expected for 1963), but with respect to the increasing length of the articles. Various measures were discussed to meet the problem, including a page charge.

Also in this meeting, the 'vulnerability' of BBA with only two Managing Editors was discussed. Replacements during vacations, illness or a long visit abroad (Slater was again planning to visit Australia in the summer) were necessary. A 'third man' was suggested, and Remarque suggested Van Deenen, but Westenbrink was then reluctant to burden him with editorial duties at this stage of his career. Both he and Slater were also reluctant to undertake the 'training' of a third man at this moment. They had an excellent relationship with one another, could still just manage the work and were apprehensive about the idea of running the journal by a 'committee'. Geelen made the interesting suggestion that as third man an 'expert', rather than a biochemical authority, was necessary, probably in the form of an Editorial Secretary, stationed in Elsevier (see also p. 60).

At the next meeting held also in Utrecht on October 17, 1963, the situation during the previous summer was discussed. Although Slater had been able to carry on in Melbourne, it was clear that such a situation should not be allowed to arise again. Geelen had already helped in a few cases, and it was decided that he should be trained as Editorial Secretary, working first with Slater.

The question of a third Managing Editor was again discussed in detail. Slater was now more in favour, and it was decided that if it proved that an Editorial Secretariat could substantially take away the load from the Managing Editor, Van Deenen could be invited as the 'third man'.

Also at this meeting, it was decided that, at the beginning of 1964, the Advisory Board would be abolished and the members would be invited to become Editors. As a result of this decision, some very distinguished biochemists, such as H.A. Krebs, were 'shanghaied' onto the Editorial Board and did sterling service.

The question of a 'third man' was again a major item at the third meeting on February 6, 1964, the last to be held in Westenbrink's office. Westenbrink announced that he wished to retire when he was 65 (2 years away) and Slater said that he would also probably wish to stop in about 4 years, since he was considering accepting an offer in Australia to take over the Melbourne chair in 3–4 years.

Slater was still not happy with the sectionalization. Since most articles on biological oxidation were published in the section Enzymology, he suggested changing the name to 'Enzymology and Biological Oxidations'. Westenbrink made the important suggestion to move papers in photosynthesis from Biophysics to the same section as biological oxidations, since photosynthesis was becoming more and more biochemical. This suggestion led ultimately to the Section 'Bioenergetics', one of the most successful in the journal, particularly with respect to the articles on photosynthesis.

REPLACEMENT FOR WESTENBRINK

In the spring of 1964, Westenbrink fell ill, necessitating hospitalization. He telephoned Slater asking him to take over full responsibility for BBA while he was in hospital, which he expected to be for only a short period. On June 10, 1964 a meeting was held in Slater's office in Amsterdam. Slater announced that he had decided not to accept the offer in Australia. Various possible constructions for the editorial organization of BBA were now discussed: (a) present structure; (b) decentralization with one Managing Editor for each section; (c) regional editors; (d) centralization, with one editorial bureau and one or more Editors-in-Chief.

After a detailed discussion, which is extensively recorded in the minutes of the meeting, it was agreed that possibility (d) was the only feasible one at that time, although Slater liked the idea of possibility (b) in principle. The minutes are in Dutch and citations have been translated.

> "JR asks ECS: 'Would you like to be Editor-in-Chief'?
> ECS answered that this depends on the future organization. He cannot spend any more time than at present. He could not even for a short period (a half year) do any more and could not devote more time for building up a secretariat. During the present situation, ECS is prepared in difficult cases to help JG who is at the moment doing Westenbrink's work.
>
> JR asks ECS: 'Do you agree with the construction: Editor-in-Chief with an assistant or associate Editor-in-Chief and an extensive secretariat'. ECS answered that by elimination of constructions (a), (b) and (c) no other possibility remains. He emphasizes once again that a good secretariat is the most important part of the structure: better a good secretariat and a bad Editor-in-Chief than vice versa.
>
> The assistant Editor-in-Chief is now discussed. There are a number of good biochemists in the Netherlands, but who is ready to undertake the task? There are no emeriti, and also none in the next ten years. As suitable persons Prof. Van Deenen and Prof. Gruber are mentioned."

It was decided that Remarque would lay the plan before Westenbrink and if the latter agreed, Remarque would inform the Editorial Board, after which a definite decision would be taken.

A meeting planned for 29 June did not take place, probably because it had not been possible for Remarque to see Westenbrink. In view of the urgency of the matter it was decided to invite Van Deenen and Gruber to be Associate Managing Editors, and to put into immediate operation the concept of an Editorial Secretariat. Jan Geelen was put in charge of the Secretariat, which consisted at first of Mrs. Koning, who had been with Westenbrink since 1959, and Miss Tilanus, who had been with Slater since 1961.

On 29 July, during the Sixth International Congress of Biochemistry, a meeting was held in New York of the Editorial Board.

This was attended by 22 editors, together with Bergmans, Remarque, Klijn and Mr. B. Russak, President of the American Elsevier Publishing Company. Bergmans announced the death of Westenbrink two days earlier and that Slater was willing to become Managing Editor.

Slater explained the new organization. Apart from the Managing Editor (this term had now replaced Editor-in-Chief), there would be two or more Associate Managing Editors, and Van Deenen and Gruber had agreed in principle to accept this function. A BBA Editorial Secretariat had been formed, manned by three trained biochemists, one English. The secretariat would make routine decisions regarding acceptance or refusal of manuscripts submitted and carry out correspondence with the authors, signing letters as Secretary of Editorial Board on behalf of Slater. (In fact it was decided later that letters would be signed by the Editorial Secretariat, acting on behalf of the Managing Editors.)

The new organization centering around the Editorial Secretariat and Managing Editors was explained in detail. Little has been changed since then and the minutes of the New York meeting could still serve as a guideline to those newly joining the secretariat.

The summer of 1964 was extremely difficult for BBA. In the absence of the Managing Editors, the Editorial Secretariat were thrown into the water before they had learned to swim. Inevitably, mistakes were made, but Slater decided that the show must go on, and that the Editorial Secretariat must take decisions on the basis of reports from editors and referees. However, he insisted that on his return from the summer meetings he would review all decisions and all letters to authors so that mistakes could be pointed out and avoided in the future. In this way, a working relationship between the (Associate) Managing Editors and the Editorial Secretariat was devised, whereby the Secretariat became the executive arm of the Managing Editors. The following policy was laid down, which is still adhered to:

(1) If the reviewers (Editor or referee) are clearly in agreement (ac-

cept or reject), the Editorial Secretariat puts the decision into effect without further consulting a(n) (Associate) Managing Editor.

(2) If the two reports do not agree, if after a reasonable length of time (now one month) only one report has been received, or if more than minor points for improvement are raised by the reviewers, the matter is brought before a(n) (Associate) Managing Editor for decision.

(3) No paper may be rejected purely on the basis of reports of referees, that is, unless at least one editor (or (Associate) Managing Editor) has been consulted.

Copies of all correspondence with the authors were to be sent to all (Associate) Managing Editors, so that, where necessary and possible, a decision could be modified or, more often, the Secretariat was instructed to deal with such matters in the future in another way or word a letter differently.

This retrospective control is in fact a major and time-consuming task of (Associate) Managing Editors, especially since the staff of the Secretariat changes quite frequently. Twice, a head of the Secretariat left Elsevier to join the prestigious journal *Nature,* and one of them later became Manager of Elsevier's Cambridge office.

The meeting of the three Managing Editors (Slater, Van Deenen and Gruber) with Remarque, Geelen and Kleijn on September 7, 1964, set the pattern of these meetings which has also not changed much in the next 20 years. Recurring points of discussion are (re)-invitations to editors, publication delays, delays in correspondence with authors, number of volumes to be published in the next year, suggestions and instructions to authors, complaints from authors and readers, 'case histories'. In the first years, the enormous growth of BBA, which will be discussed in the next chapter, was a matter of concern. Other specific items will also be dealt with in the appropriate chapter. Special attention was given in the early meetings to a Westenbrink memorial issue. In the first issue of 1965 (Volume 97), the first 18 pages were devoted to biographical notes concerning Westenbrink written by colleagues and fellow editors, and included some very fine photographs.

The meetings were at first held monthly, from October 1964 in the Elsevier office. Occasionally, if a member of the Editorial

Board happened to be in Amsterdam, he was also invited. At the November, 1964, meeting, Mrs. Koning and Miss Tilanus were present for the first time. Dr. J.H. Morris, the first English-speaking biochemist to join the Secretariat, was at the February, 1965, meeting. As the Editorial Secretariat gained more experience and procedures were clearly laid down, the frequency of meetings could be diminished until now only two are held each year. The increasing 'professionalism' of the Secretariat and also of Elsevier's sub-editing staff becomes clear as one reads the minutes. Morris started producing regular statistics of number of papers received and accepted and the publication times in different sections.

On February 3, 1966, the Managing Editors were invited to a luncheon with the Board of Elsevier's Scientific Publishing Company to celebrate the publication of the 100th volume of BBA. The Editors were surprised to see that the 'faceless' Board included, in fact, some colleagues from the University world in the Netherlands and Belgium.

On September 7, 1966, Slater called a private meeting with Van Deenen and Gruber to discuss the establishment of the new FEBS journal *(European Journal of Biochemistry)* and also the question of appointing an additional Associate Managing Editor, for the first time from outside the Netherlands. As a result, the experienced biochemist and editor Professor Albert Neuberger joined the team at the end of 1966.

INFORMATION EXCHANGE GROUPS

At the December, 1966, meeting attention was given to the problems created for the journals by the Information Exchange Groups established by the U.S. Public Health Service. The meeting endorsed a statement prepared by the IUB Commission of Editors of Biochemical Journals which is set out below.

"The Commission of Editors recognizes the value of the Information Exchange Groups as a medium for rapid exchange of informal suggestions, comments, queries, criticisms and general discussion among groups of scientists who share a common interest in a particular field, provided that such memoranda are not intended for publication; this was indeed the original purpose of the I.E.G.'s and the Commission hopes that they will continue to serve this purpose. In order to make this purpose clear the Commission recommends that each I.E.G. memorandum should clearly state on its front page that the memorandum is not intended for publication and is not to be quoted in published papers.

On occasion some authors, after circulating new findings or ideas informally in an I.E.G. memorandum, will later incorporate the same material – usually with considerable modification – in a paper submitted to a journal. We would regard this as a proper and normal procedure.

However, the circulation of an I.E.G. memorandum that is identical (or nearly identical) with a paper simultaneously submitted for publication in a journal can cause much trouble and confusion. The paper may undergo drastic revision before acceptance by the journal; in that case many workers in the field will read the earlier unrevised version, and may fail to read the published paper. The confusion that has arisen in some cases is serious and most unfortunate.

Moreover there are grave objections to the circulation of manuscripts already accepted by journals as I.E.G. memoranda; that is, the distribution of preprints by an agency entirely independent of the publisher of the scientific paper. This involves great added expense and raises serious questions concerning possible violation of copyright.

In view of these considerations the Commission of Editors proposes that its member journals adopt the following policies:

(1) No paper will be considered for publication if that paper, in essentially the same form, has previously been released as an I.E.G. memo. Papers may not be submitted simultaneously to a journal and to I.E.G., nor may papers already accepted for publication in a journal be released through I.E.G.

(2) I.E.G. memoranda are not to be cited as such in the published paper. An author may refer to the information contained in such a memorandum as a 'personal communication' from the writer of the memorandum. This requires the written consent of the writer and the Editor may require evidence of this before permitting the inclusion of reference to such a personal communication."

QUALITY OF PAPERS

Naturally, the question of the quality of papers published in BBA was a constant topic of discussion. At the meeting on March 14, 1967, Remarque put the situation clearly. If one were to apply a scale of quality evaluation to all biochemical papers submitted to all biochemical journals, with values ranging from 1 (for the very best 'top' articles) to 10 (for quite worthless 'trash' papers), then the situation of BBA could be described schematically as follows:

> Papers of quality 10 and 9 are not or very seldom submitted to BBA and, if so, are rejected. Papers of quality 8 and 7 are being submitted more frequently to BBA but for the greater part are also rejected. Some of the papers of quality 7 and most of the quality 6 are accepted for BBA after extensive revision, improving their quality and resulting in a quality of 6 or 5, respectively, after accepting them. In the same way most of the articles of quality 5 are 'refined' to quality 4 before accepting. The bulk of papers submitted to BBA is of a quality 3 or 4. However, 'top' quality papers in the groups 1 and 2 are rarely submitted to BBA. The general opinion was that *The Journal of Biological Chemistry* and *Journal of Molecular Biology* were publishing more top papers than BBA and that *Proceedings of the National Academy of Science, U.S.A.* now had many first-rate articles.

The situation today is probably not very different. The average citation frequency of an article in BBA is below that of *Journal of Biological Chemistry* or *Biochemistry,* but is of the same order as that of other 'general' biochemical journals (see Table II). However, the quality varies greatly from Section to Section. Although no statistics are available, it can be confidantly predicted that the citation frequency of the Bioenergetics and the Biomembranes sections is much higher than average, and that of Nucleic Acids (now Gene Structure and Expression) lower.

The quality of the latter section has been of continuing concern to BBA. Undoubtedly, the *Journal of Molecular Biology* and *Proceedings of the National Academy of Science, U.S.A.* were taking many of the good papers in this field. It is not, however, immediately

Table II

Number of citations and citation impact of leading "general" biochemical journals in 1983

Journal	Number of citations*	Citation impact**
Arch. Biochem. Biophys.	18,794	2.44
Biochem. Biophys. Res. Commun.	35,458	3.02
Biochemistry	45,234	3.84
Biochem. J.	39,266	3.43
Biochim. Biophys. Acta	69,210	2.54
Eur. J. Biochem.	26,076	3.40
FEBS Lett.	23,860	3.00
J. Biol. Chem.	141,620	6.11

* Number of citations in 1983 to all volumes published since inception.
** Number of citations in 1983 divided by numbers of articles published in 1981 and 1982.

obvious why BBA quite suddenly lost so much ground. Once lost, it proved very difficult to regain. Even the appointment of additional Managing Editors (C. Weissman, R.J. Flavell and P. Borst), all very distinguished in this field, for this section was not successful in this respect, although they were a very welcome strengthening of the policy-making body in general. This section has continued to lose ground to the smaller specialist journals, including *Gene,* founded by Elsevier in 1976, as part of an expansionist policy.

In 1979, Elsevier sent a questionnaire to BBA's editors, referees and readers, who were asked their opinion about its functioning and how it could be improved. More than 750 replies were received and extensively analysed by Elsevier. The results of this survey were discussed by the Managing Editors in their meeting on January, 1980. It confirmed that the main problem concerned the section on Nucleic Acids and Protein Synthesis. Slater commented that there is no doubt that BBA had badly 'missed the boat' in this field. He recalled a meeting many years earlier of the Editorial

Board of the *Journal of Biological Chemistry* to which he had been invited, in which the editors complained that their journal did not receive the exciting papers in the newly developing fields of molecular biology – they went to *Journal of Molecular Biology* and BBA. *Now,* BBA and the *Journal of Molecular Biology* are in the same position as the *Journal of Biological Chemistry* was then, and the latter has recovered.

Biochemists, particularly those in molecular biology, do like to follow the fashion, and this changes rapidly. The fashionable journal in their field was once BBA, then it became *Journal of Molecular Biology* then the *Proceedings of the National Academy of Sciences, U.S.A.,* now *Cell.*

NOTES TO CONTRIBUTORS

During the discussion between Slater and Gaade about nomenclature referred to in the previous chapter, it became clear that it would be desirable to issue a guide to contributors to BBA. Gaade, who was experienced in questions of chemical nomenclature, undertook to prepare a list of symbols and abbreviations, that was at first intended only for in-house use. The only mention of nomenclature in the first 'Notice to Contributors' in 1958, which was printed on the inside cover, was the following paragraph.

> "In principle, *Biochimica et Biophysica Acta* follows the nomenclature and symbols adopted by such international bodies as the International Union of Pure and Applied Chemistry, and the International Union of Biochemistry. Pending adoption of a uniform system by one of these international bodies, the procedure of the *Biochemical Journal (Biochem. J.* 66 (1957) 6) will be followed in referring to isotopically labelled organic compounds. Abbreviations for chemical substances should be used sparingly, and should follow the principles adopted by the *Biochemical Journal (Biochem. J.* 66 (1957) 8), and the *Journal of Biological Chemistry (J. Biol. Chem.* 233 (1958) 3). All abbreviations used should be defined in a statement which will be printed at the bottom of the first page of the paper."

In a memorandum dated June 3, 1960, Remarque commented that during a visit to Czechoslovakia he had received the suggestion that BBA publish an editorial on "How to prepare a contribution to BBA". He hoped that this would reduce the sub-editorial work and, at the same time, serve as a 'primer' for young authors.

This task was undertaken by Slater, in consultation with the Editorial Board, and eventually appeared in 1961 as a leaflet, available free of charge, with the name "Suggestions and Instructions to Authors". Although grateful use was made of similar 'Instructions' by other biochemical journals, notably the *Biochemical Journal,* the BBA "Suggestions" went further into the way a paper should be written. It was particularly addressed to the large number of BBA authors with less experience in writing papers, and it became a 'best seller'. Slater's students were instructed to write the report of their research project in accordance with the BBA Suggestions and Instructions (except that these reports were in Dutch).

Already in the 1961 'Suggestions and Instructions', BBA followed enthusiastically the recommendations of the IUPAC (later IUPAC-IUB) Commission on Biochemical Nomenclature, although in view of the strong feelings at first against using NAD instead of DPN, the following phrase appears:

> "For the time being, the editors will not insist that the recommendations on the nomenclature of coenzymes will be followed."

In the second edition (1965) this was replaced by:

> "The Editorial Board of *Biochimica et Biophysica Acta* strongly recommends that authors follow this internationally accepted nomenclature. On the other hand, the editors recognize that certain authors feel strongly, as a matter of principle, that the older DPN-TPN nomenclature should be retained. In these cases, the editors will permit the use of the DPN-TPN nomenclature."

This rider was dropped in the third edition (1969).

Jan Geelen collaborated closely with the IUB Committee on the

Nomenclature of Enzymes to help familarize authors with the new nomenclature (see p. 59).

Documents prepared by the Commission of Biochemical Nomenclature were rapidly printed in BBA and, in accordance with an agreement made with other members of the IUB Commission of Editors of Biochemical Journals, 'clean pulls' were made available for reproduction in other biochemical journals. Later, however, resistance arose to using valuable journal space for printing long documents dealing with chemical nomenclature with only marginal interest for biochemists.

BBA's "Suggestions and Instructions" was recommended by the *European Journal of Biochemistry* to its authors and the fourth edition (1971), renamed "Information for Contributors", was prepared in collaboration with Professor Claude Liébecq, Editor-in-Chief of that journal. This was reprinted, with minor amendments, in 1976. A fifth edition appeared in 1979 and a sixth in 1982. In addition to being supplied as a brochure, this edition was printed in the journal [1].

Where at one time BBA was in the vanguard of the campaign to encourage the use of good nomenclature, the enthusiasm of the Managing Editors to do so has waned considerably in recent years. As a reflection perhaps of the general questioning of authority in the 1970's, the working biochemist became more and more impatient at being told how he should write. In this period also an Office of Biochemical Nomenclature was set up in the United States with the laudable aim of helping authors to follow good nomenclature. Unfortunately, it did this by sending photocopies of published papers to the Editorial Boards and to authors heavily scored with comments about nomenclature, and sometimes even about the style of writing. Although the points made were nearly always well taken, they irritated the recipients and Editorial Offices. The BBA Editorial Secretariat, in particular, was overwhelmed by a stream of such missives and, although Slater urged the members of the Secretariat not to take them as criticism but as helpful guidelines, there is no doubt that the well-meaning and indeed in many ways useful efforts of the Office of Biochemical Nomenclature made the Managing Editors less receptive to recommendations of the IUB-IUPAC Commission of Biochemical Nomenclature.

However, there is no doubt that, although the use of unfamiliar nomenclature and capital-letter abbreviations has increased rather than decreased in recent years, the bad old days of the 1950's, when incorrect and even misleading nomenclature and the unnecessary use of capital-letter abbreviations were widespread, are now much less prevalent. Some 'bad' abbreviations like SDS have become almost part of the language, but Slater still winces when he sees SDS-PAGE.

SECTIONALIZATION OF BBA

The sectionalization of BBA, that was started with individual special issues in *'Nucleic Acids and Related Subjects'* and *'Enzymological Subjects'*, was rapidly developed in the next few years, with frequent changes of names (see Fig. 3). By 1967, however, the sections became stabilized and, with the exception of the introduction of *Reviews on Biomembranes* (1972), *Reviews on Bioenergetics* (1973) and *Reviews on Cancer* (1974), remained unchanged until 1982. In that year the sections *Protein Structure* and *Enzymology* were combined under the name *Protein Structure and Molecular Enzymology* the section *Nucleic Acids and Protein Synthesis* was renamed *Gene Structure and Expression* and a new section, *Molecular Cell Research*, was started. However, these developments belong to a later chapter.

It is now time to return to the 1960's and consider the growth explosion of BBA that dominated its second decade.

REFERENCE

1 Information for Contributors to Biochimica et Biophysica Acta (1982) *Biochim. Biophys. Acta* 715, 1–23

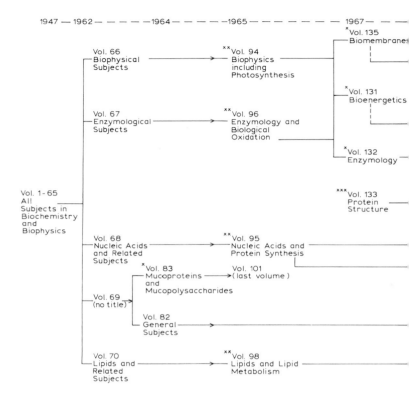

Figure 3. Growth and differentiation of BBA. *Specialized sections formed from less specialized sections. **Volume title changed. ***New section. Issues 69/3, 74/2 and 78/2: specialized issues on mucoproteins and mucopolysaccharides. Also published: Cumulative Subject Index, Volumes 1–50, 101–149; Combined Author and Subject Indexes, Volumes 1–50 and yearly since 1968; Author Indexes for Volumes 51–100, 101–149. (Figure prepared by C. Pollard.)

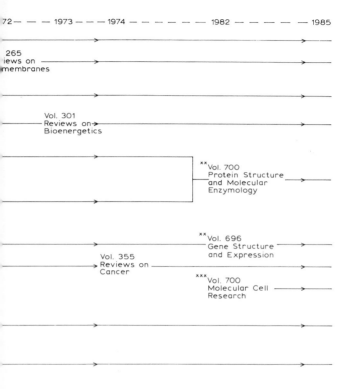

APPENDIX – BIRTH OF BBA SECRETARIAT

by J. Geelen

In the autumn of 1962 Westenbrink approached me at a cocktail party in his laboratory* saying: 'I heard that you are not so happy with your present job and I thought you might be the right person for BBA at Elsevier's".

Later, I found out that the vacancy at Elsevier was partly due to the desire of Slater and Westenbrink to implement the recommen-

---

* At that time I was working at the Rijks-instituut voor de Volksgezondheid after my specialization in physiological chemistry at Westenbrink's institute.

dations of the Enzyme Nomenclature Commissions in BBA*. Thus, even before my start in the new job I received from Elizabeth Steyn Parvé the brand new first edition of Webb's 'bible' to prepare myself.

On a cold, snowy 3 January 1963, the well-known 'heart of Africa' incident happened again: two young men shook hands in the corridor of the as yet unfinished Elsevier office in the Jan van Galenstraat saying: "Ko Klijn/Jan Geelen I presume"**. After a crash course on the same day by Remarque, he took us the following day to a printing house; we were then plunged into the BBA room, only to find that the chief of the BBA desk-editorial team, van Dijk, had resigned in order to finish his chemistry studies, that his relative, Erna van Dijk, had left simultaneously, and that the only person left was desk-editor Hans van der Tweel. Notwithstanding the frustration of not being promoted to run BBA, he gave us a very thorough desk-editorial training and until Ko had employed our first secretary there was no time at all to replace 'old yellow enzyme' by its new name and add its EC No.; typing of contents, reading proofs and composing issues were the first priorities.

On Thursday 9 May, 1963 the first Managing Editors' meeting was held in the laboratory of Westenbrink, which I attended. Westenbrink and Slater brought up the heavy workload and the vulnerable position of editorial management of papers, especially in view of the imminent departure of Slater to Australia for a couple of months. Remarque suggested appointing Van Deenen as a third Managing Editor. At this stage of the meeting I made the suggestion that not just a particular authority but rather an 'Editorial Secretary' might be the solution in the future. On our way back to Amsterdam, Remarque was furious: "How can you dream of

---

* The implementation of the recommendations of the EC in BBA was no sinecure. I had to spend at least an hour daily to weed out old or wrong names, try to identify others and finally to write standard letters to authors seeking their approval or resolution of ambiguities. (1260 papers were screened and 245 letters written in 1963.)

** We had heard about each other from a mutual acquaintance.

making suggestions at a meeting without previous consultation – do you realize that, if they take your suggestion seriously, it will cost Elsevier a fortune!"

In August 1963, Westenbrink asked me to come twice weekly to Utrecht to be initiated into the editorial work. There I sat with Lida Koning, his BBA secretary on Elsevier's payroll, in a corner of the library suggesting editors for refereeing, suggesting decisions and drafting acceptance/revision letters. In the meantime, Slater had arranged to continue his work while in Australia, but the workload proved to be very heavy, postal delays were frustrating and finally his BBA secretary, Marijke Tilanus, fell ill. On 10 September he phoned me from Australia: "This must never happen again; I would like to discuss this extensively in terms of the last May meeting when I am back". Immediately after Slater's return it was decided that I should be trained by him as well. Thus, from now on I worked another two days a week in a mini-corner of his laboratory on the Jonas Daniel Meijerplein and only one day at Elsevier.

In May 1964 the situation changed dramatically. Westenbrink fell ill and asked me to take over all his work, i.e., taking complete responsibility, including refusal of papers and dealing with ambiguous cases. This situation gave rise to intensive brainstorming between Slater and Elsevier. Both agreed on the principle of an Editorial Secretariat for the Managing Editor, independent of, but on the payroll and in the office of, Elsevier. Meticulous care was taken to give credence to this fact of independence through, amongst other things, the introduction of a separate post-office box and maintenance of anonymity. I remember penetrating sermons from Remarque never to disclose my identity and become 'Mr. BBA' and from Slater, warning me not to grasp authority as seems to have happened with other publisher-owned journals or to get fussy over details of presentation like some society-owned journals.

Against this background, an Editorial Board meeting during the IUB Congress in New York was being organized for July 29 at which the Editors would be informed and consulted about the new Editorial Secretariat. On July 27 Westenbrink passed away.

On 14 August, soon after the return of Slater, now joined by Van Deenen and Gruber, the Secretariat was born with all its procedures formalized and implemented, thus securing unbiased double refereeing, decisions in ambiguous cases exclusively by Managing Editors, and strict anonymity.

The first problem for the new-born Secretariat was rooming and personnel. My pendulum movement between the laboratories was not very practical; Lida had to move out of Westenbrink's laboratory, and Marijke, in the unbelievably overcrowded laboratory of Slater, but not working for the University, had to move as well. The amount of work under the rigid rules was much too much for me alone and my knowledge of the English* language desperately needed some support.

On January 1, 1965 John Morris joined the Secretariat. He was already in Holland working in post-doctoral research, but was more interested in publishing. He applied for a job in Elsevier just at the right time, and he stayed until mid 1969**, when he had to return to England for family reasons.

On 1 February 1965, the Secretariat settled down in a large room in Elsevier's premises and officially began its existence in the eyes of the world. Lida moved happily from Utrecht, but Marijke stayed only for a short period, since she was not willing to forego the laboratory atmosphere.

---

* In November/December I had to abandon BBA for one month to serve my Queen. Fortunately, this war game was played in England, where I had for the first time the opportunity to observe in its natural surroundings what would become the dominating species in the Secretariat's population!

** John Morris has been admired by everybody for his modesty, but more than this, his work was so good that when he expressed his intention to leave, a 'London-satellite' BBA Secretariat was seriously contemplated. Procedures were being worked out and accommodation was being sought. However, the consensus was eventually reached that any solution centred around one person, however good that person might be, would be too vulnerable, and it was very fortunate for John that he could find similar work at the MRC in London.

CHAPTER 5

The growth explosion of BBA in the 1960's

THE 1966 'STATEMENT OF POLICY'

At the first of the regular meetings between the Managing Editors (then Westenbrink and Slater) and Elsevier (Remarque, Geelen, Klijn) on 9 May, 1963, Slater expressed his fear concerning the enormous growth of BBA. For the June, 1965, meeting he prepared a memorandum which was the basis of 'A statement of policy by the Managing Editors of *Biochimica et Biophysica Acta*' that was published in 1966 [1] after extensive consultation with the Editorial Board (no less than 20 replies were received to a circular, an unusually high response) and with the IUB Commission of Editors of Biochemical Journals.

Fig. 4, which is reproduced from this statement of policy, shows that, between 1947 and 1965, BBA had in fact shown a uniformally logarithmic growth with a doubling time of 4 years. It was only during the 1960's that the consequences of the logarithmic growth were making themselves felt. In particular, the Managing Editors

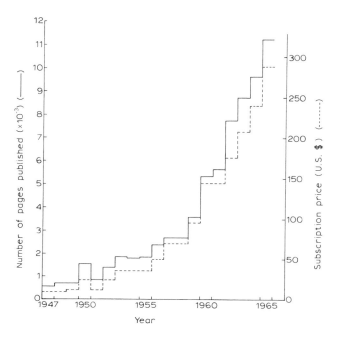

Figure 4. Volume and subscription price of BBA between 1947 and 1965 (reproduced from Ref. 1).

were receiving complaints from their colleagues concerning the escalation of the price. As one colleague put it in a letter in November, 1964:

> "I first subscribed to it, when it was in the range of $46.00 a year, but it soon got too rich for my blood and we have been picking it up with funds we obtain from some of our various sundry friends for use in this function. Now we have gotten to the place where it looks to me like the Journal is going to take the majority of our funds. I just got the notice a while back that next year's subscription will be $288.00. When I look at the bulk of science and the quality and nature of the publication, I wonder if this is really an accurate measurement of either compared to the *Journal of Biological Chemistry* which comes to members at $22.50 and to the general public at $45.00. Would it be impertinent to ask what is the basic reason for this enormous cost, what do you project for the journal in terms of size and cost, and

particularly per page of science cost? I fear that you will soon price yourself out of our market and many others too. Is there anything that we could mention to you which would be useful or is this a commercial publication, handled by Elsevier, with you folks merely doing the handwork on the editing."

Fig. 5, also taken from the Statement of Policy, showed that, as expected, the growth of BBA was only part of the total growth of the biochemical literature. However, in the period 1950–1965, the

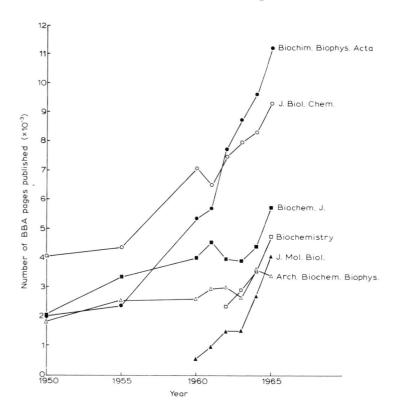

Figure 5. Volume of six major biochemical journals between 1950 and 1965 (reproduced from Ref. 1).

total number of papers in 12 journals surveyed had increased by 289% compared with 466% for BBA. Thus BBA's share had also increased during this period, and in 1962 BBA had surpassed the *Journal of Biological Chemistry* as the largest biochemical journal. It was also noteworthy that, apart from BBA (27%), most of the increase in the biochemical literature in the previous 5 years had been absorbed by new journals (*Biochemistry* (22%), *Journal of Molecular Biology* (16%) and *Biochemical and Biophysical Research Communications* (5%)).

A simple calculation showed that a doubling time of 4 years, if unchecked, implied that 36 volumes would have to be published in 1969 and 72 in 1973. Indeed, one could further extrapolate that by 1983, BBA would be publishing 512 volumes per year, say one every morning and one every afternoon in a 5-day working week.

The straight-from-the shoulder punch in the letter cited above concerning the cost of BBA and similar verbal comments from their colleagues caused the Managing Editors great concern. In this respect, the interests of the Managing Editors and the Publisher do not run completely parallel. Both have the same aim – to provide a service to the biochemical community – but they have different responsibilities. The Managing Editors are responsible, morally if not legally, to their biochemical colleagues. The Publisher is responsible to his shareholders. Both parties are concerned that the journal should be read, and therefore that it should be easily available to biochemists at a price as low as possible. However, the Managing Editors recognize that Elsevier, like any other business, has to make a profit and that if BBA shows a loss, it will have to stop publication, which, considering the large percentage of the world's biochemical publications that have appeared in it, particularly in the 1960's, would be a disaster. Moreover, as already illustrated in Chapter 3, a new journal inevitably makes a loss that has to be recouped in later years. Also, as in any other industry, innovations (in this case new journals) and services (such as the publication of symposium proceedings at a loss) can be financed only by profits from existing ventures. The Managing Editors were not, however, altogether happy that a part of the proceeds from the sales of BBA went to founding new journals that competed with BBA for good papers.

The Managing Editors had no insight into the magnitude of the profit made by the Publishers from BBA, so could not directly assess whether it was 'reasonable' or not. This was a deliberate policy on their part. Both Managing Editors and Publisher agreed on a strict division of responsibility. Editorial matters, in particular the selection of papers, is strictly a matter for the Managing Editors, and financial matters for the Publisher. However, in view of the comments received by the Managing Editors, they had to satisfy themselves that the price of BBA was not unreasonably high.

It was obvious that, in view of the considerations listed above, the price of BBA would be much higher than that of the well-established journals owned by scientific societies. How much higher was not so obvious. All that the Managing Editors could do was to compare it with that of other commercially operated journals.

The subscription price was plotted in the 'statement of policy' (see Fig. 4), but not further commented upon. In Slater's memorandum for the June, 1965, meeting, an additional graph was given showing the price per BBA page from 1947 to 1964, in comparison with the average for U.S. commercial journals and for U.S. society-owned journals. In 1961, the latest year for which the latter information was available, the values ($US per 10,000 words) were: BBA, 0.36; U.S. commercial journals, 0.39; U.S. society journals, 0.17. This information was not included in the printed 'statement of policy' by the Managing Editors, but in November a confidential memorandum was sent to the Members of the Editorial Board with comparative prices of journals. For society journals, this varied from $0.51 (*Journal of Biological Chemistry*) per 100 BBA pages to $1.64 and for commercially published journals $1.78 to $3.94. BBA at $2.56 was about average among the latter. This memorandum, which is reproduced below, was also sent to the members of the IUB Commission of Editors of Biochemical Journals on December 3, 1965. It was widely read, with particular interest by those involved in the Federation of European Biochemical Societies (FEBS).

CONFIDENTIAL
BIOCHIMICA ET BIOPHYSICA ACTA
MEMORANDUM
from: Managing Editors and Editorial Secretariat
to: Members of Editorial Board

Price of BBA
The Publishers and the Managing Editors receive frequent complaints about the price of BBA. Fig. 1 of the Memorandum on the problems posed by the growth of biochemical literature shows that the price per million words has not, in fact, changed since 1957, during a period of great wages and price increases in The Netherlands. Since 1957, a number of costly improvements have been made without increase in price. In 1966, the very important step of delivery by airmail to U.S.A., Canada, Japan, Australia and New Zealand will be undertaken. For U.S.A. and Canada, the only extra cost is a handling charge of $0.10 per issue. For Australia, Japan and New Zealand the opportunity is given to receive the journal by airmail for the cost of $1.00 an issue, which is, in fact much less than the cost.

The actual price of BBA is compared with that of some other journals in the Table. The journals are grouped according to whether they are owned and operated by a scientific society or by a publisher.

It is clear that BBA is not exceptionally dear, in comparison with other publisher journals. The comparison is often made with *J. Biol. Chem.* It would appear to us that the correct conclusion is that *J. Biol. Chem.* is exceptionally cheap, rather than that BBA is exceptionally dear.

One might add that an average-sized paper in *Biochemistry* (7.1 pages) costs almost as much in page charges as a subscription to BBA (7.1 × $35 = $248).

Amsterdam
November 1965

The Managing Editors were pleased to note that the $288 subscription referred to by their colleague represented, in fact, about the same price per BBA page as 10 years earlier.

This was a fine managerial achievement on the part of the Publisher, in view of the wages explosion in the Netherlands in the early 1960's. Unfortunately, it was the total subscription price that had to be found within the libraries' budgets. It was the growth of BBA rather than the price per volume that was the problem.

The last paragraph of the November 1965 memorandum notes that one of the society journals operated a page charge and that the cost to the author of publishing an average-size paper in this journal was about the same as a year's subscription to BBA. The Managing Editors and Publisher have always agreed in opposing both page and handling charges for BBA. In addition to the general objection that it transfers the financial burden from the reader, where it belongs, to the writer, there was the practical and indeed moral objection to charging authors from the economically less advanced countries, who also publish in BBA.

The minutes of the June 1965 meeting make interesting reading today. There was a good and frank discussion between the Managing Editors and the Publisher, during which Bergmans clearly expressed the realities of commercial publishing. All agreed that it would be desirable to set a ceiling at 18 volumes for BBA, but no clear decision was taken how this should be done. Bergmans said that it was clear that the biochemical literature would continue to grow, and that if a ceiling were put on BBA, other journals would be created. Elsevier, like other publishers, would like to participate in this growth and he had in mind creating an additional biochemical journal. The Managing Editors were not too happy about this. Although Slater would have preferred two journals of 18 volumes each rather than one of 36 volumes, he foresaw difficulties about having any direct connection – at the editorial level – between the two journals. The decision to 'freeze' BBA to 18 volumes was announced by Slater at a meeting of the IUB Commission of Editors of Biochemical Journals in Paris on 7 July, 1965.

At the next meeting in August of Managing Editors and Publishers, it was recognized by all present that the decision absolutely to freeze BBA to 18 volumes was too drastic and it was decided to publish 19 volumes in 1966. This was also to be maintained in 1967. This was announced in the 'Statement of Policy' [1] which also specified further steps in an attempt to restrict the growth of BBA.

> "The very rapid growth of biochemical literature is, most likely, simply caused by a corresponding increase in the number of biochemists. This means that every biochemist has more and more to read. Since the working day remains of the same length, the proportion of the world's biochemical literature that he reads is declining. He makes use of other methods of keeping himself informed of developments in his own field, for example by attending specialist symposia or through an information-exchange group where one exists in his field. As a consequence, he tends to know less and less about developments in fields outside his own particular interest. This is particularly regrettable in a subject such as biochemistry, where developments in one field may have a direct impact on another.
>
> As in other branches of science, a communication crisis is developing in biochemistry. This will no doubt be solved eventually by making use of recent technical advances, but the nature of the solution is at present difficult to foresee. In the meantime the biochemist will continue to report his work in scientific journals. He should realize, however, that it is in his own interest, because he wants his papers to be read, to write as concisely as possible. The average biochemical paper is often longer than those customary in many other branches of chemistry.
>
> Conciseness should not be achieved, of course, by omitting details essential for the understanding of the work or for repeating it. Nor should reference to preceding work on the subject be omitted. Much can be done at the paragraph level simply by removing repetitions. The greatest saving in space can be brought about by restricting the paper to the most significant results achieved in the research. The essential features of confirmatory or control experiments can often be given in a few sentences in the text.
>
> For some time now, the Editorial Board has been giving special consideration to the length of papers, and many requests have been made to authors drastically to shorten their papers, even to the extent

of resubmitting the paper as a Short Communication. In our opinion, this policy has not only relieved the pressure on space in the journal but has greatly improved the readability of the papers and, thereby, increased their impact on the reader."

An important result of the discussion in the June 1965 meeting about the price was Elsevier's decision to send BBA airmail to Canada and U.S. free of extra charge, except for a small handling charge, and subscribers in Australia, New Zealand and Japan were offered airmail service at a modest extra charge. BBA was the first journal to make use of airmail post on such a scale.

FAILURE TO CONTAIN THE GROWTH OF BBA

In the course of 1967, it became apparent that the Publisher wished greatly to expand BBA in 1968. At the meeting held on March 14, 1967, Remarque put forward the following plan for discussion (quotation from minutes of meeting):

> 'In order to attract more high-quality papers, BBA should aim at a constant, competitive low publication time for all sections. Until recently, only retrospective steps were taken to correct the publication time when it threatened to become dangerously high. Remarque proposed a publication policy wherin the publication time (preferably low) is considered to be the constant starting point. Publication within 4 months after acceptance can be guaranteed if the Publisher is allowed to adopt the size of the issues and the volumes in a flexible way."

In the discussion that followed, Remarque and Slater clashed. To quote the minutes:

> "Slater pointed out that the size and price of a journal fall within the editorial responsibility. Remarque disagreed."

Further discussion was deferred to an extra meeting held a month later on April 18, 1967. At this meeting, Slater expressed the following objections to Remarque's proposals:
(1) we should be going back on our decision, published in the 'Statement of Policy', to restrict the growth of BBA;
(2) the Editorial Board would lose control of size and price;
(3) the present pressure on Editors to advise rejection or condensation of papers would be reduced. Although Remarque would guarantee no increase in subscription for 1967, Slater was afraid that this plan would lead to a drastic increase by 1970. Slater felt particularly that the influence of the Board on the Journal is of concern to the biochemical community and that BBA's image could be adversely affected by an extreme publisher-owned journal policy. Goodwill and the capacity to appeal to the best workers, such as the *European Journal of Biochemistry* was exhibiting, is vital. In his opinion, the adoption of Remarque's proposal would lead to a loss of any control by the Editor over the size and price of BBA. The Managing Editors were being asked to sign a blank cheque for 3 years. Remarque answered that the Directors of Elsevier required freedom in principle in order to be able to plan effectively.

Finally, the Managing Editors agreed to a flexible schedule under continual review. They would be expected to be consulted on the price for 1968 and in later years. This expectation has been fulfilled. During the Spring meeting each year, the size and price of BBA for the subsequent year are decided by mutual consultation.

So far as 1968 is concerned, it was agreed that two more volumes would be published in 1968 with the price at $16 per volume, as it had been since 1958. This was reported to a meeting of the Editorial Board, attended by 20 Editors, during the International Congress of Biochemistry in Tokyo, on August 23, 1967.

Sixteen years after the first editorial, a second one appeared in BBA [2] recalling and reaffirming the two paragraphs cited from the 1966 'Statement'. In this second editorial, the graph reproduced in Fig. 6 was published, showing that, just as in biological

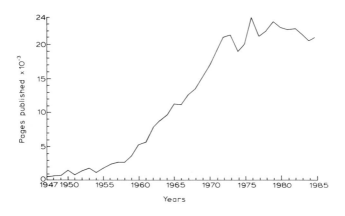

Figure 6. Volume of BBA between 1947 and 1985 (reproduced from Ref. 2, with updating).

systems, the growth of BBA had slowed down. The steps that were announced in 1966 probably had only a marginal effect, although the efforts to persuade authors to write more concisely did meet with some success. Articles in BBA are on the average considerably shorter than those in comparable journals. The main reasons for the slowing down in the growth of BBA are probably: (1) a slowing down in the growth of the number of biochemists; (2) the establishment of a few other major biochemical journals but, more important, of a large number of specialist journals in biochemistry and biophysics. In 1965, the total number of pages published by 12 major journals listed in the 'statement of policy' (about 50,000 BBA pages) was a reasonable measure of the world biochemical literature. In 1980, the number of pages in the major biochemical journals (*European Journal of Biochemistry* had replaced *Biochemische Zeitschrift*, and *FEBS Letters* had been added to the 1966 list) had rather more than doubled (to about 110,000 BBA pages) compared with 1965, just as BBA had, but the total growth of biochemical literature was probably considerably greater. Accurate data on this last point are, however, not available.

A complete explanation of the phenomenon described by Fig. 6

is not available. What are the factors that have limited the growth of BBA? Other major biochemical journals have followed qualitatively a similar course. This is an interesting subject for research.

In the event, Elsevier did not start a new journal covering the whole field of biochemistry and biophysics. The discussions initiated by Slater in the IUB Commission of Editors of Biochemical Journals were probably not without their effect on the decision by the FEBS to found two new journals, the first in 1967 when the *European Journal of Biochemistry* replaced the long-established *Biochemische Zeitschrift* so that the number of biochemical journals was not increased. Relations between the new journal and BBA have always been cordial. For some time, Slater was Chairman of the IUB Commission of Editors of Biochemical Journals and Professor Claude Liébecq, the first Editor-in-Chief and still Chairman of the Managing Editors of the *European Journal of Biochemistry*, was Secretary and later succeeded Slater as Chairman when the latter became Treasurer to IUB. For the first years, the new journal used BBA's Notice to Contributors, with the latter's blessing. The *European Journal of Biochemistry*, although only about one-third the size of BBA, has certainly been instrumental in restricting the growth of the latter.

In line with their opinion expressed during the discussions around the possibility of Elsevier establishing a new journal, both Elsevier and the Managing Editors of BBA considered it undesirable that Elsevier should publish the second journal founded by FEBS, *FEBS Letters*, and in the event this was undertaken by the North-Holland Publishing Company. In 1970 the latter company fused with Elsevier, and both *BBA* and *FEBS Letters* are managed, from the publishing side, by the Biomedical Division of Elsevier Science Publishers, B.V., a subsidiary of Elsevier N.V. They remain, however, completely separate entities and there are no connections whatsoever between the two journals at the editorial or editorial secretariat level.

DATA DEPOSITION

As one method of trying to restrict the growth, BBA, in common with other journals, from 1971 encouraged authors to deposit supplementary data (after editorial review) with a data-deposition agency and offered one of their own. As a method of reducing the size of BBA, this had only a marginal effect. Between 1971 and December 1985, 243 data sets were deposited with Elsevier and 261 requests had been received for the data from 165 data depositions. Thus 78 sets had been stored without any requests for the data having been received.

The great majority of data sets relate to papers in the protein and nucleic acid sections with details of determinations of amino acid and nucleotide sequences. Some journals now publish such data (as well as details of the experimental methods used) in mini-print.

BBA's experience with data deposition schemes is not encouraging. Perhaps new developments in computer science will provide the answer.

REFERENCES

1 Anon (1966) *Biochim. Biophys. Acta* 121, 223–227
2 Editorial (1982) *Biochim. Biophys. Acta* 714, 1–5

CHAPTER 6

The problems of the 1970's

THE NUMBER OF SUBSCRIBERS AND THE PRICE OF BBA

The number of subscribers to the complete BBA package (including full set equivalents from the sections) reached a peak of 2,800 in 1968. Since that date, in common with many of the other larger biomedical journals, a steady decline set in (see Fig. 7).

The increasing diversity and specialization of interests in biochemistry was an important factor in the eventual decision to make BBA also available in separate sections. The introduction of such individual sections considerably improved the subscription situation, and the total resulting circulation showed a far less dramatic decline (see Fig. 7).

The decline in the journal's circulation seen since 1968 can be attributed to a number of factors including the increasing specialization in science generally, the general economic decline (especially in many Western European countries) in the late 70's; the advent of considerable 'resource-sharing' by libraries and library networks leading to the erosion especially of second and third subscriptions

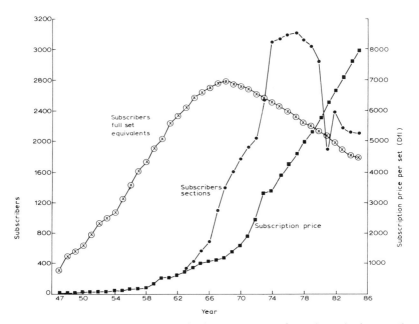

Figure 7. Subscriptions to BBA and subscription price from the early days until 1985. The number of subscribers refers to full set equivalents, viz. the number of subscribers to full sets *plus* the number of subscribers to the sections when converted to full subscriptions.

leading to the erosion especially of second and third subscriptions to BBA on many university campuses, the wide-spread introduction of photocopying machines, the increasing use of 'secondary' retrieval services by which potential readers of BBA could locate and retrieve relevant articles through third-party sources, competition from other biochemistry journals especially the large number launched in the 1970's, and last but not least the considerable growth in the size of BBA itself that had occurred in previous years and the resulting price increases, which were also fuelled by inflationary pressures.

In Fig. 8, the price of the total set is shown together with its two components – the price per page and the number of pages per set. Also plotted is the theoretical price per page taking the same start-

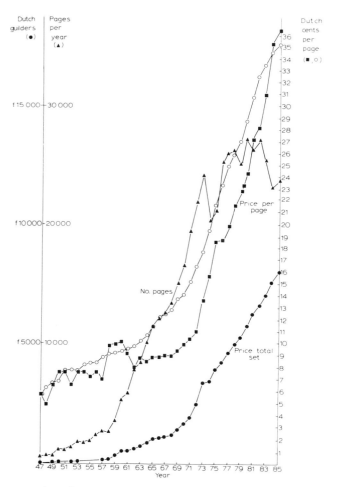

Figure 8. Number of pages (▲), price of total set (●) actual price per page (■) and theoretical price per page, calculated as the 1947 price adjusted for changes in the purchasing value of the guilder (○) for BBA between 1947 and 1985. Prices in Dutch guilders.

ing point in 1947 applying the cumulative effect of inflation in the Netherlands between this date and 1985 (source: Centraal Bureau voor de Statistiek). This shows that until the early 1970's the actual price per page moved very little despite a 60% decline in the purchasing value of the guilder between 1947 and 1970. After 1972, the price per page increased steeply following the fierce inflation of the 1970's. In 1985, it was the same as the 1947 price adjusted for inflation.

Thus the efforts made to contain the price in the middle 1960's were not without effect. In 1972, the Managing Editors were still concerned at the continuing growth in the size of the journal. During the FEBS Meeting in Amsterdam in the summer of 1972, a large number of editors accepted the invitation of the Publisher to a beautiful dinner at a seventeenth century villa outside Amsterdam (24 August 1972), in order to celebrate the 25th birthday of BBA. Ter Haar announced that, as a 25th birthday present, Elsevier was donating 25,000 guilders to IUB to be used to assist young biochemists and biophysicists to attend international meetings. Slater, who was now Treasurer of the Union, accepted this gift with another hat on his head. All editors received in addition as a present a fine book with reproductions of the work of Ortelius, one of the early Dutch cartographers.

In his after-dinner speech, Slater made a new extrapolation of the future of BBA.

> "This evening we are celebrating the 25th Anniversary of the founding of BBA. A 25th Anniversary always has a special significance, especially in the Netherlands, where we like to use every fifth anniversary as the occasion for a party – called a lustrum. A lustrum of lustrums – 5^2 – is of course something very special and will not be repeated for another 100 years, when it will be 5^3. None of us will be around then. Will BBA be?
>
> Some of you may remember that some years ago the Managing Editors and the Editorial Secretariat undertook an essay in futurology, and even found a journal (BBA of course) ready to publish it. This is brought up-to-date every year by Mr. Geelen, and distributed within the IUB Commission of Editors of Biochemical Journals, under the title 'The growth of the biochemical literature'.
>
> A simple extrapolation of the data of BBA for its first 25 years into the next 100 years yields fascinating results. Whereas in 1972 one issue of BBA appears every 2 or 3 days, 100 years from now five issues

will appear every second – or 1 every 200 ms, or 18,000 issues every hour. If we make the reasonable assumption that the size of the Editorial Board increases in proportion to the journal, simple mathematics leads to the inescapable conclusion that 100 years from now, this meeting will be attended by 100 million editors.

This will create some technical difficulties but I am sure that our Editorial Secretariat – expanded to roughly 5 million – will solve them. In any case, I am sure that the long-term planning department of Elsevier, advised by Professor Böttcher and his colleagues in the Club of Rome, are working on this.

Another problem, however, will probably have to be dealt with by the Editorial Board, because it is molecular biological. It seems doubtful if the world will contain in 2072 a sufficient mass of nuclear DNA with a base sequence suitable for an editor of BBA. Professor Theorell, speaking on behalf of IUB at the opening ceremony of the FEBS Meeting, has pointed the way for a solution. Professor Theorell predicted that, within 10 years, we should be greeting synthetic colleagues. Yesterday evening I was reminded by the Rector of the University of Amsterdam, a learned man and a biologist, that man likes to create in his own image. The solution is at hand. Elsevier will create a store of Charles Weissmanns and Piet Borsts, synthesized according to the procedure described in a BBA report, and fill up the Editorial Board as required. In this way, we need have no fear of the I.Q. of the Editorial Board declining – on the converse, it might rise quite alarmingly.

Of course, extrapolation into uncharted terrain covering 4-times the area of the charted terrain is unjustified, although some Lineweaver and Burk plots appear in BBA in which this is done. However, the serious point that I wish to make is that all I have done is to extrapolate known data, and that the growth factor that I have used for this calculation *has* appertained in the last 25 years. I have chosen to extend the line forwards instead of backwards in order to bring home to you just what has been achieved by BBA since it was founded by the late Professor Westenbrink and by Mr. Bergmans.

The founding of BBA was due to Westenbrink's vision in forseeing that the advance of science in Europe – freed from the restrictions of War – would require new journals, in particular one for biochemistry and the new science of biophysics, and to Elsevier's resolve to back him. 25 years later BBA is objectively the largest biochemical journal and maybe, for this occasion, and quite subjectively of course, we can say one of the best.

This result has been achieved only by the Publishers and Editors (and in particular the Managing Editors) working close together –

each responsible in the sphere in which they have special expertise. Of decisive importance for the development of BBA was the setting up, in 1964, of an Editorial Secretariat. Apart from dealing with the enormous amount of normal secretariat work, the Editorial Secretariat has formed a most effective link between the Editorial Board and the Publisher, having as they have responsibilities to both bodies. The Publisher has never interfered – or even made suggestions – with respect to editorial decisions. I must confess that I cannot say the reverse – the Managing Editors have been quite vocal with respect to decisions belonging to the sphere of the Publisher, although recognizing that the final decision belonged to the latter.

Publication is an essential part of scientific research – the free communication of his findings to his colleagues is, in fact, the primary social responsibility of the scientist supported by public funds – if I may refer to a competing occasion this evening. [A meeting on the social responsibility of scientists – ECS.] Editorial boards and publishers play, then, an essential role in the furtherance of scientific research, and it is the recognition of this fact, I believe, that explains why so many leading scientists are ready to spend so many of their evenings and week-ends doing editorial work. Without this selfless willingness to help their colleagues (in a way that is not always appreciated) and the development of their subject, it would not be possible to continue a journal like BBA. It is fitting that on an occasion like this, we acknowledge this. It is also good that we do not take this willingness for granted, especially at a time when the traditions of the academic world are changing rapidly."

In making the extrapolation beyond 1972, maybe Slater's voice carried too far. In any case, in the following year the first oil crisis hastened the end of the apparently unrestricted growth of many activities in the West, including BBA. Moreover, the rapid increase in the rate of inflation and the large changes in currency rates soon took the question of the price of BBA out of the sphere where the Managing Editors could make an effective contribution. Whereas in 1965 the price of BBA was in the middle range of that of commercially owned journals, the changed value of the guilder in relation to that of the US dollar had as consequence that by 1973 BBA's dollar price was now considerably higher than that of other comparable commercially owned journals. Moreover, the gap between the price of some society and commercially owned journals had widened, partly at least due to the introduction of page charges by the former. As already mentioned on p. 69, both Publisher and Managing Editors firmly opposed the introduction of page charges in BBA. In the first half of the 1980's, the doubling of value of the US dollar in relation to the guilder has led to a substantial decline in the cost of the subscription to BBA in US dollars.

THE EDITORIAL SECRETARIAT

After the disastrous summer of 1964, the Editorial Secretariat, under the leadership of Jan Geelen and John Morris, worked out procedures which became a model for those used by secretariats set up later by other journals.

During the dinner in the Netherlands to celebrate the 25th Anniversary of the founding of BBA, one Editor suggested that he and his fellow Board Members could become better informed about the central organization of the journal through the medium of a regular newsletter. The first appeared in December 1972 with the optimistic intention of issuing three or four Newsletters per year. In fact, only 13 have appeared in the subsequent 12 years.

The first Newsletter gave an instructive flow diagram of the editorial handling of manuscripts, which is reproduced in Fig. 9.

Unfortunately, a rapid turnover of staff and, at times, serious under-staffing at a time when the number of papers to be processed was still growing rapidly led to difficulties and to tensions, in which the Editorial Secretariat became the football to be kicked between the Managing Editors and Publisher*.

During the 1970's the Managing Editors became increasingly concerned that the Editorial Secretariat was slipping away from its

---

*As a member of the Secretariat recently aptly remarked, without a ball there is no game.

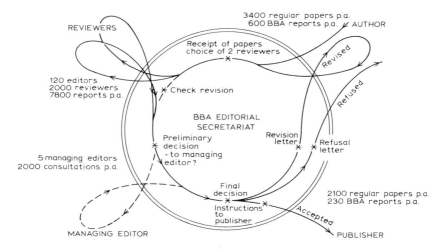

Figure 9. A flow diagram of the editorial handling of manuscripts, with figures from 1971 (from Newsletter No. 1; figure designed by J. Geelen).

control. In 1974, the Secretariat had, on its own initiative, sent policy statements to the members of the Editorial Board, laying down criteria for rejecting papers. The Managing Editors agreed with the comments of some of the editors that a too rigorous following of the criteria could lead to BBA becoming a rather stereotyped and narrow journal and that, in its attempts to eliminate dull papers, it would itself become a 'dull' journal even if all the papers that it published were competent. Accordingly, a letter signed by both Slater and Dr. A.M. Grimwade, the then Head of the Secretariat, was sent on October 30, 1974, to all Editors endorsing this concern. The letter added . . .

> "For this reason and also because the list of reasons for *not* accepting a paper inevitably creates a negative impression, they are strongly of the opinion that these policy statements should not be made available to authors. Although it is possible to lay down negative points, it is more difficult to express on paper why a paper is interesting, exciting, stimulating, in other words good.

> The inevitably subjective decision of an editor to accept or refuse a paper is based on a number of tangible and intangible factors. . . .
> In general, however, we still want you to make your own judgement. The general editorial policy of the journal, which is to accept papers that report significant and novel advances in biochemistry and biophysics, remains unchanged."

Towards the end of the 1970's the publication time gradually increased. Partly due to staff shortages in the Editorial Secretariat in the summer of 1979, the total publication time for this year was between 28.4 weeks (Enzymology) and 34.5 weeks (Nucleic Acids). More serious were the numerous cases of excessive delays in informing authors of acceptance or refusal of their papers. For some reason or other, mechanisms that had previously been developed to signal that a decision was being delayed too long by the non-receipt of a referee's report were no longer working. Indeed, for a time it seemed that only a telephone call from a harassed (would-be) author could reactivate a dormant paper. There was no doubt in Slater's mind at least that what became known among the Managing Editors as the 'disastrous summer of 1979' lost a lot of goodwill for BBA among biochemists.

How the problem of publication delays was tackled is a matter for the next chapter.

The delicate question of the relationship of the Editorial Secretariat to the Managing Editor on one side and the Publisher on the other came to a head again during Slater's last year as Chairman of the Board of Managing Editors, as the result of a series of misunderstandings around two hand-written memoranda concerning 'refusal letters' that Slater had sent to the Editorial Secretariat.

As he stated in a letter to the Managing Editors discussing the episode.

> "The primary task of an Editorial Secretariat is to act as Secretariat to the Editorial Board. As Chairman of the Editorial Board, I have the right to expect the Editorial Secretariat to carry out my instructions when they are purely secretarial, without question. If they concern editorial matters, they have the right (and indeed the duty) to dis-

agree and to discuss the matter with me. Therefore, I had no objection to their memorandum as such. I did object and would never have agreed to sending it to you before I had discussed it with them. In fact, this was done in good faith because, owing to a misunderstanding, they had understood that I had given permission.

In any case, by sending out their memorandum, the Editorial Secretariat had, in my opinion, taken up a position that does not belong to them. BBA is run by the Publisher and the Editorial Board, *assisted* by the Editorial Secretariat. The precise relationship of the Secretariat to the Publisher and the Editorial Borad has, for far too long, been insufficiently clear. What is, however, *now* crystal clear and always should have been, is that the Editorial Secretariat is not a *third* party, on a par with the Editorial Board and the Publisher."

The matter that brought this to a head was apparently trivial, but in fact had its origin in a feeling of frustration by the Managing Editors that, whereas they would like to be sympathetic and helpful to authors, the scale of BBA made a more bureaucratic approach inevitable. BBA had moved a long way from Westenbrink's ideal, so well worded by Wyckoff (see p. 21).

The diagram reproduced in Fig. 9 shows that at the beginning of the 1970's, the Managing Editors were directly concerned with only about one half of the papers. This low fraction was justified by virtue of the retrospective control referred to on p. 49, and the existence of a body of experience with respect to the nature of the reports of individual editors that had accumulated within the Secretariat in the late 1960's. Indeed, Slater and the Secretariat applied unofficial 'correction factors' to the reports of certain editors. As the result of rapid changes in the composition both of the Secretariat and Editorial Board that took place in the 1970's, however, the retrospective control became less effective, and the Managing Editors are now involved in more and more papers – about 75% in 1985.

RELATIONS BETWEEN MANAGING EDITORS AND PUBLISHER

It will be clear to the reader of the previous chapter that tensions developed between the Managing Editors and the Publisher as the result of the problems arising out of the growth explosion. This was a time, however, when, particularly in continental Europe and perhaps even more particularly in the Netherlands, the accountability of those in authority (particularly the Professors!) was a burning question. The Managing Editors felt themselves morally accountable to the biochemical community, but did not know how to render this account. In the final analysis, the Publisher had complete control and could dismiss and appoint Editors at his will.

At the first meeting in which the worries of the Managing Editor concerning the price of BBA were expressed (on May 12, 1965), Slater stated that he supported the idea of a sponsorship or a curatorium above the Publisher, since there was much criticism in the scientific world about publishing with an eye to making a profit.

After the turmoils of 1965–1967 and the realization that the attempt to keep the lid on the kettle of BBA had failed, a good working relationship was built up between Publisher and the Managing Editors. At the end of 1968, however, new tensions developed. On March 15, 1969, Slater felt obliged to write to his colleagues – at that time Van Deenen, Gruber, Neuberger and Weismann – to express his concern at the relationships with Elsevier, and particularly with respect to the way the periodic meetings were developing.

In Slater's view the meetings were primarily of the Managing Editors with the Secretariat, attended also by the Publisher. It was, after all, on that basis that he chaired the meetings. Furthermore, it was clearly understood that the Editorial Secretariat should work under the direct supervision of the Managing Editors. In the last months, however, decisions were being announced by the Publisher that concerned directly the Managing Editors, without previous consultation with them. Thus, at the November, 1968, meeting the Publisher announced the appointment of a new member (a biochemical graduate) of the Secretariat chosen without any consultation with the Managing Editors. At the opening of the meeting on March 12 the Publisher announced that both Remarque and Morris were leaving the company, the latter to return to England, the former to join the Elsevier Holding Company. The pending departure of Morris from the Editorial Secretariat was known to the Managing Editors; that of Remarque was a bolt out of the blue. His departure was a serious blow to the Managing Editors. He had been a tower of strength for BBA for more than 10 years and, despite tensions at times, a good working relationship had been built up. Morris, who had together with Geelen got the Secretariat off the ground, would also be sorely missed.

Slater's letter of March 15 was discussed with the Associate Managing Editors at a meeting in his laboratory held immediately before the next regular meeting.

In Slater's notes for the meeting, which are still in his archives, can be read

> "1. Must distinguish between discourtesy, which we can grin and bear, and actions which we believe are incompatible with our editorial function. 2. We must consider the impact that our (or my) resignation could have. 3. In any case, we cannot break all contacts with Remarque so abruptly, without some sign from our side of our appreciation for what he has done."

As a result of this meeting, two letters were sent by Slater, one of thanks to Remarque and inviting him to dine with the Managing Editors, and a second to the Publisher. The second letter requested a formalization of the relations between Managing Editors, Editorial Secretariat and Publisher to be laid down in a Letter of Intent, and stated that in the opinion of the Managing Editors, the relationship between the Managing Editors and the Publisher need not differ in principle from that between a professional society and the Editorial Board of its journal. The Publisher agreed to this request in a formal letter on August 11, 1969, from Bergmans writing as Managing Director of Elsevier Publishing Company which stated:

> "Following the pleasant exchange of thoughts on July 1, 1969 concerning the need to formalize the relationship between the Editorial Board of BBA, in particular the Managing and Associate Managing Editors, and the Publisher, I am happy to confirm herewith the points we agreed upon:
>
> 1. Responsibility for the scientific assessment of manuscripts submitted rests entirely with the Editorial Board, in particular with the Managing Editors.
>
> 2. The Managing Editors are appointed by the Publishers on the recommendation of the existing Managing Editors.
>
> 3. The members of the Editorial Board are appointed in consultation between the Managing Editors and the Publisher. The aim is to have a board in which the various branches of biochemistry are as fully represented as possible and the geographical distribution is as favourable as possible.
>
> 4. Decisions regarding matters of general importance such as financial matters, numbers of volumes, frequency of appearance, and subscription prices are taken by the Publisher after consultation with the Managing Editors.
>
> 5. The BBA Editorial Secretariat deals with all matters concerning manuscripts submitted to the Journal and is responsible for this to the Managing Editor. With regard to all other matters its responsibility is to the Publisher.
>
> 6. The staff of the Editorial Secretariat is employed by the Publisher and is appointed by the Publisher after consultation with the Managing Editor.
>
> I hope that his letter of intent covers all points of our discussion, and I look forward to receiving your confirmation."

This agreement certainly cleared the air and cordial relations developed, particularly with O. ter Haar, who became the fixed personification of Elsevier at a time when the direct leadership of BBA from the Publisher's side was rapidly changing. The transfer of Remarque marked the end of an important era in the history of BBA, and the Managing Editors found it difficult to adjust to the rapidly changing faces of the representatives of the Publisher. They sometimes had the impression that each new appointee had to make his or her 'mark' on the firm by making some change in BBA, necessary or not. At the meeting on February 20, 1975 a new face was seen and a new voice was heard – Dr. Jack Franklin, who became a worthy successor to Remarque.

The general question of the relationship between the biochemical community and publisher-owned journals was again raised by Slater at the close of a meeting on September 7, 1971, in connection with comments that had in fact been made relating to another journal, not published by Elsevier/North-Holland. The poor standard of this journal was taken by some as an example of irresponsible action by commercial publishers in sponsoring journals with a non-representative editorial board. Slater concluded that

> "The feeling that scientific publishing should be controlled by scientists instead of publishers is growing, and commercial journals would have to try and identify themselves with the biochemical public in some way."

He mentioned that the Pan-American Association of Biochemical Societies (PAABS) had produced a memorandum in which it was stated that all biochemical publishing should be run by an organization set up by the biochemical societies.

One response to a circular to the Members of the Editorial Board concerning criteria for rejection (see p. 84) included the following:

> "I suppose the only kind of information that Elsevier does not share with us is the profit – or loss – margin at which they operate. If we take it that the break-even point with 2,000 subscriptions is the 1973

> page level, that is fine; if, however, the suggestion is that we cut the number of publications and leave the profit margin alone, then we have some reason to wish to participate a little more deeply in the proposition proposed."

For reasons given earlier, the Managing Editors did not wish to participate deeply in the question of the profit margin for the Publisher.

The next chapter in this story of the relationship between publisher and editor arose out of the fact that Slater was Treasurer and, therefore, member of the Executive Committee of IUB. By that time, FEBS was obtaining a substantial income from its two successful journals, one (*The European Journal of Biochemistry*) in collaboration with Springer, the second (*FEBS Letters*) in collaboration with North-Holland Publishing Co., which in 1969 had fused with Elsevier. Dr. W.J. Whelan, who had been the leading figure in FEBS circles behind the founding of the two journals and was now General Secretary of IUB sought a similar income for the Union.

Slater consulted his colleague Managing Editors in a letter written on September 30, 1973. He pointed out the difficulty caused by his double position – BBA's Managing Editor and member of the IUB Executive Committee – but added.

> "In any case, I propose to examine the matter purely from the viewpoint of BBA and, if it comes to negotiations, I shall sit on the BBA side of the table.
>   What are the advantages to BBA of an official link with IUB? These are, I believe:
> (1) An improvement of 'BBA's image' in approaching (the structure of) a society-run journal, rather than being a purely commercially run journal.
> (2) It would give a mechanism by which the biochemical community could participate in important decisions, such as those we have been discussing recently and in choosing our successors as Managing Editors.
> (3) It would make it easier for the biochemical community to resist developments not in the interest of biochemists.
>   What are the disadvantages?

(1) The Managing Editors might feel inhibited by an IUB Publication Committee 'breathing over its shoulder'.
(2) Decisions might be made more democratically, but by less expert 'votes'.
(3) A flexible response to a new situation would be (made) more difficult by an increase in the number of consultations required.
(4) It might well increase the burden on the Managing Editor to such an extent that the task can no longer be combined with an active biochemical job.
(5) Political considerations and self interest might influence policy decisions.

Some at least of these disadvantages could be eliminated by a suitable agreement protecting the Editorial Board and the Publisher from undue interference. So far as I know, Liébecq is able to operate efficiently as Editor-in-Chief." [Professor Liébecq was Editor-in-Chief of the *European Journal of Biochemistry*.]

Since Elsevier could see only disadvantages and no advantage to a link with IUB, they could not agree to the latter's proposal. However, they brought forward a counter-proposal. For some time, they had been considering bringing out a new journal, *Trends in Biochemical Sciences* (TIBS), aimed at providing the individual biochemist with easily accessible and highly readable short reviews, highlighting the best recent literature and currents in research in specific subfields of biochemistry. In a meeting with Whelan and Slater in Elsevier's office on December 17, 1973, Elsevier offered IUB sponsorship of this journal and this was accepted.

Thus, once again BBA was indirectly involved in the foundation of new journals. In 1965, its publicly announced policy to try to limit its growth was very likely a factor, although perhaps not an important one, leading to the foundation of the two FEBS journals. IUB's approach to Elsevier concerning BBA undoubtedly encouraged Elsevier to launch TIBS in collaboration with IUB. The first miniature (and free) issue of TIBS appeared in October 1975 with as caption 'Published for the International Union of Biochemistry by Elsevier'. Whelan as first Editor-in-Chief of the journal set the pattern for its continued success, the number of subscribers reaching 7,000 in 1985. Unlike BBA, the size of TIBS is restricted by the aim to keep its price within the range payable by the individual biochemist. Nevertheless, the royalties are now making a significant contribution to IUB's income. TIBS was the first of a whole family of 'Trends' journals now published by Elsevier.

The issue of the relationship between the Publisher and the Managing Editor was raised once again in a "relatively dramatic" form (as he put it) by Franklin, who on April 1, 1980, wrote to Slater stating that the Publisher's side was increasingly worried about the increasing age of the Editorial Board and the fact that BBA had lost, over the past few years, its image as an innovative journal. He stated that, in order to focus the individual sections upon their markets, he would like to "promote" the individual Associate Managing Editors to Managing Editors of their section, and to strengthen the Board by appointing new Managing Editors to cover BBA's less strong areas. He proposed that certain Managing Editors step down, that Slater become Chairman of the Editorial Board, remaining in overall control until his 65th birthday and then step down to "just running" the Bioenergetics section, a position that he would hold while he remained in full and active research.

With the Publisher's permission, Slater consulted the two Managing Editors who were not expected to 'step down' (van Deenen and Radda) on May 13, 1980, and a meeting was held with the Publisher on June 17th, in which as Slater put it:

> "There were six little nigger boys, now there are two. We are facing the biggest change in BBA for about 20 years, and we need to think pretty carefully where we are going."

Slater added that he was planning to retire from his Chair in the summer of 1985.

No further discussions took place until January 15, 1981, when the Managing Editors met alone before the main meeting. Slater made no secret of his personal unhappiness at Franklin's letter, but more important in his view was how the episode pointed out the power that the Publisher had with respect to the largest biochemical journal in the world. He mentioned that for more than 10 years he had been trying without success to find a structure in which there is a buffer between the Publisher and the Editorial Board as there is in all society-run journals. He added that, in theory, the Editorial Board has a tremendously strong card to play against the Publisher, since without editors they are powerless. However, it is in practice very difficult to play this card. It is like doctors going on strike, or Professors refusing to mark exams. How could we face our biochemical colleagues, if we went on strike? If BBA did not appear for a month, there would be a terrible dislocation of scientific publication.

It was decided to negotiate a contract with Elsevier in which the responsibilities, obligations and rights of the Managing Editors and the Publishers were laid down. The contract, which was signed on January 1, 1982, has as its main items:

> 1. The copyright, name, and ownership of the journal remain as the exclusive property of the Publisher.
>
> 2. The Board of Managing Editors (hereinafter called 'Editor') will be responsible for the scientific control of BBA and undertake to maintain the highest possible scientific standards. Any proposed changes to the scientific policies of the journal will be agreed upon by both parties. The Publisher agrees to inform the Editor of matters which impact on his responsibility, e.g., publication of new journals which may compete with BBA.
>
> 3. The Editor will, in consultation with the Publisher, appoint new members to the Board of Managing Editors in situations of resignation, expansion or other agreed reasons. The Board of Managing Editors will, in consultation with the Publisher, select a member of the Board to be Chairman. The Chairman will be appointed for a period of 3 years, and may not normally serve more than 2 consecutive terms of office. The first person to be appointed Chairman is Professor E.C. Slater, who will serve in this capacity until December 31st, 1982. The Chairman will be selected by the Editors at least 12 months prior to the resignation of the incumbent Chairman.

4. The Editors will select an Editorial Board to assist them in their duties. The Editorial Board will be chosen so as to cover the different fields and scientific communities represented by BBA.

5. The Editors will select one or more additional editor(s) who will be responsible for the sections, 'Reviews on Biomembranes', 'Reviews on Bioenergetics' and 'Reviews on Cancer'. Appointments will be made following consultations with the Publisher.

6. The journal will publish original papers and review papers and such reports as may be decided by the Editors. All papers submitted will be critically evaluated according to scientific criteria to be set by the editor. The Publisher will assist in the evaluation process by providing, at the Publisher's expense, an Editorial Secretariat which will work under the scientific control of the Editor, or, if the Editor elects to receive manuscripts directly, secretarial assistance, to handle the administrative processes involved in the publication of BBA. The Editorial Secretariat will be appointed by the Publisher in consultation with the Editor. Otherwise, the Editor will be responsible for appointing secretarial staff to his own office.

7. The format, layout, number of pages to be published, and pricing of the journal, are the responsibility of the Publisher. The Publisher will consult with the Editor on these issues.

8. This Agreement is valid for a period of five years from the date of signing and will automatically be renewed for further periods of five years unless one year's notice to the contrary is given by either party.

9. All disputes arising in connection with the present contract, or the breach thereof, shall be finally settled by arbitration in the Netherlands under the rules of the Netherlands Arbitration Institute. Judgement upon the award rendered may be entered in any court having jurisdiction thereof.

After further discussions, Slater agreed to be Chairman of the Board of Managing Editors until December 1, 1982, and to remain as Managing Editor with special responsibilities for Bioenergetics until the end of 1985. From the beginning of 1982, M.K.W. Wikström and P.L. Dutton took over as Managing Editors for Bioenergetics Reviews (which Slater had previously also edited) and on January 1, 1985, Professor K. van Dam joined Slater as Managing Editor for the Bioenergetics section and took over completely on January 1, 1986, when Slater was made Honorary Managing Editor.

Thus, the Publisher achieved his aim of rejuvenating the Board of Managing Editors (see next chapter) and Slater achieved his aim of having the rights and responsibilities of the Managing Editor written into a contract. Of particular importance is that the contract is with the Board of Managing Editors and that the Board appoints new members and also its Chairman. Although Slater's idea of a curatorium has not been realized, owing to the lack of a suitable body to act as one, experience has proved that the Board of Managing Editors can be entrusted with choosing its successors.

In his farewell speech to Neuberger on December 2, 1981, Slater expressed some of his feelings about recent events in the following words:

> "Now after 20 years as Editor and 15 years as Managing Editor, BBA will have to continue without the benefit of your tremendous experience of biochemistry, editing biochemical journals and BBA in particular. You are making way for the 'young', as we all have to do eventually, whatever it may mean. Because old and young are not exact scientific terms. One's age is an objective fact, 'younger' and 'older' also have an exact significance. But whether one is young or old is a subjective phenomenon, dependent upon both the observer and the observed, and particularly on the objective age gap between the observer and the observed. Those still in their 20's or 30's may feel that anyone above 60 is old, those aged or approaching 60 will probably think otherwise."

As a result of these events Slater proposed that, in future, meetings should be regularly held in three parts, first with the Managing Editors alone, the second together with the Publisher and the third the normal 'plenary' meeting with the Editorial Secretariat, as had in fact been done at the January 1981 meeting. At the opening of the first meeting in the new format on October 14, 1981, Slater commended this structure to his colleagues. He suggested that the first part be confined to discussing matters where it is desirable to formulate a Board standpoint for discussion with the Publisher.

Important matters were on the informal agenda for the meeting on October 14, namely the draft contract with the Publisher, the election of a new Chairman of the Board of Managing Editors to succeed Slater on January 1, 1983, appointment of a new Managing Editor, some aspects of the proposed decentralization of the BBA Secretariat and the imminent retirement of Gruber as Managing Editor. Van Deenen was unanimously appointed Chairman, and this was reported to the Publisher during the second part of the meeting, as were suggested amendments to the contract, all of which were agreed to.

In practice it is not always necessary to have three meetings, parts one and two being combined when there are no specific topics for part one.

BBA REVIEWS

An innovation of the 1970's was the introduction of BBA Reviews. This possibility was first considered at the November, 1970, meeting. The Managing Editors were hesitant about including reviews in the pages reserved for the primary publishing of new findings and, indeed, the pressure on space in the journal was still severe. It was decided, as an experiment, to introduce a new volume, with its own Editorial Board, for the publication of *Reviews on Biomembranes,* and the first issue appeared in 1972. This was followed in 1973 by BBA *Reviews on Bioenergetics* and in 1974 by BBA *Reviews on Cancer.* Van Deenen and Slater were the sole Managing Editors of the first two review sections, but for the third it was decided to go outside the team of BBA Managing Editors, and M.M. Burger was appointed as well as Weissmann. BBA *Reviews on Cancer* have, in fact, had a succession of Managing Editors: C. Weissmann (1973–79); M.M. Burger (1973–80); J. Tooze (1980); R.E. Pollack (from 1980); P.W.J. Rigby (from 1983). Van Deenen is still Managing Editor of BBA *Reviews on Biomembranes* and, as already mentioned, in 1982 Slater was replaced as Managing Editor of BBA *Reviews on Bioenergetics* by M.K.W. Wikström and P.L. Dutton.

An unusual feature of BBA Reviews is that, although most of the reviews are invited, they are subjected to the same refereeing procedure as other papers submitted to BBA. Indeed, some invited reviews have been found not acceptable for publication, even after revision.

Although it has at times proved difficult to obtain a regular flow of manuscripts to ensure publication without delay of acceptable reviews, these sections have generally been well received. To authors they offer the advantage over other review series in that the volumes are numbered consecutively with other BBA volumes, so that review articles are generally readily accessible in libraries.

In 1984, a departure was made from previous policy in that invited reviews are included in the section *Gene Structure and Expression*.

CHAPTER 7

BBA enters the 1980's and makes ready for the 1990's

RESTRUCTURING OF THE BOARD OF MANAGING EDITORS

The 1980's were entered rather 'dramatically', to use Franklin's phrase, by extensive changes in the composition of what became known in 1982 as the Board of Managing Editors. After 1964, only two additions had been made (Neuberger in 1966 and Weissmann in 1968) until the second half of the 1970's. In 1977, Slater recruited Radda to the Board.

Between 1980 and 1985, Neuberger, Weissmann and Gruber left the Board and Slater became non-active. Flavell joined in 1981, but his move from London to Cambridge, U.S.A., made it difficult for him to function effectively as Managing Editor and he left in early 1985. Cohen, Borst and Van Dam joined between 1982 and 1985 (see Fig. 10).

Thus, BBA enters the second half of the 1980's with a rejuvenated Board of Managing Editors mellowed by the experience of the Chairman (more than 20 years service) and Radda (nearly 10 years).

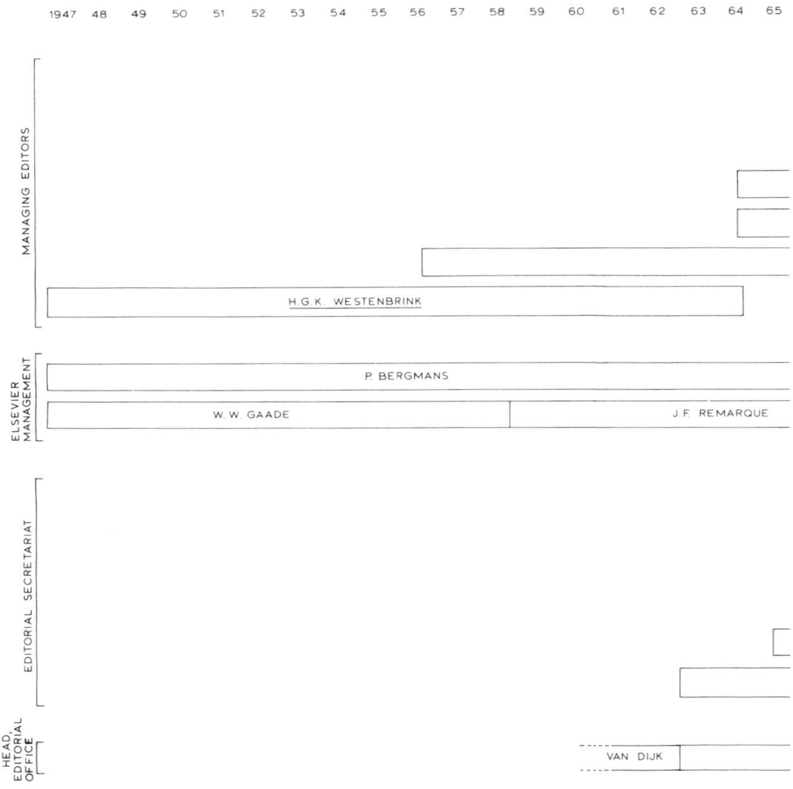

Figure 10. Involvement in BBA throughout its existence. Managing Editors (Editors-in-Chief underlined); Management responsibility (financial/policy and 'acquisition'); Editorial Secretariat (heads underlined); Editorial Office (persons responsible only).

	66	67	68	69	70	71	72	73	74	75	76	77	78	79	80	81	82	83	84	85

A. NEUBERGER — spanning 67–84
K. VAN DAM — 84–85

R.A. FLAVELL — 81–85

C. WEISSMANN — 70–81
P. COHEN — 82–85

M. GRUBER — 68–81
P. BORST — 82–85

L.L.M. VAN DEENEN — 67–85

E.C. SLATER — 67–85

G.K. RADDA — 77–85

O. TER HAAR — 67–77
J. HILLIER — 84–85

J. MEYER — 69–72
E. BOELSMA — 72–74
J. FRANKLIN — 74–77
J. FRANKLIN — 77–79
J. HILLIER — 79–83
P. JACKSON — 83–85

J.W. KOPER → 77
← P. SMALDON — 82

J. THOMPSON — 72
A.J. COLBORNE — 73–77
S.E. LORD — 77–81
H. VAN LIEMPT — 81–83
K. AUWERDA — 83–85

P. THEAKER — 70–72
H. KAMMINGA — 72–75
H.D. BERKELEY — 75–77
C. POLLARD — 77–83

A. GRIMWADE — 72–75
S.G. RICHARDSON — 75–78
N.H.C.M. BOOTS — 78–83

J.H. MORRIS — 66–69
T. HEINS (née ADAMS) — 69–74
B. MINTER — 74–77
J.A. EGAN — 77–83

J. GEELEN — 67–74
M. STREAMER — 74
K.A. LLOYD-DAVIES — 74–78
J.W. DYER — 78–83

J.G.M. KLIJN — 66–73
R.A. LUPTON — 73–79
G.R. METTAM — 79–84

RESTRUCTURING OF BBA SECTIONS

These structural changes in the Board of Managing Editors and the recent retirements and new appointments (except those of Borst and Van Dam, which came later) were announced in the Editorial published in 1982 [1].

Pending changes in the sectionalization were also announced in the following words:

> For some time, the Managing Editors had considered reorganizing and renaming the sections of BBA but had decided to await the analysis of the survey* before coming to a final decision. Starting in 1982, the following changes are being made.
>
> (i) The sections *Protein Structure* and *Enzymology* are combined under the name *Protein Structure and Molecular Enzymology*. The new name is more in keeping with developments in these fields in which we hope to attract more papers. Many papers dealing with the isolation of (new) enzymes and enzyme kinetics (except when closely related to the mechanism at the molecular level) will appear in the section *General Subjects*.
>
> (ii) It has been clear for some time that the name *Nucleic Acids and Protein Synthesis*, that dates from 1965, no longer describes the most interesting developments in this field which deal with the gene and its structure, expression and regulation. By changing the name to *Gene Structure and Expression*, we hope to attract more papers in this field than we have in recent years.
>
> (iii) A new section called *Molecular Cell Research* is being started. This section will cover papers dealing with the investigation of cell biology at the molecular level, making use of noninvasive probes of intact cells. The structure and function of isolated cells will belong to this section, as will cell secretion and uptake of material into the cell, and cell-cell interaction. The criterion will be that the results of the investigation have significantly contributed to our insight into molecular mechanisms. Biochemistry has gone through a long and successful period of reconstructing events in the intact cell from observations with broken cells and cell extracts. This knowledge must, however, eventually be integrated by studies of the intact cell in which the control mechanisms are intact. New methods of studying isolated cells or even intact organs now make this possible. BBA would like to attract good articles in this field.

---

* See p. 53.

The three sections *Bioenergetics, Biomembranes* and *Lipids and Lipid Metabolism* will continue to cover the same fields as at present. As its name implies, the section *General Subjects* covers those papers not falling into one of the specialized sections. Some fields that have until now appeared in this section may be more appropriately placed in the section *Molecular Cell Research*. On the other hand, some papers that have until now appeared in *Enzymology* will be transferred to *General Subjects*.

The new section *Molecular Cell Research*, introduced particularly at the suggestion of Radda, has been a noticeable success. Starting in 1982 with two volumes, already five are allocated for 1986. However, BBA still receives rather few outstanding articles in the fields covered by *Protein Structure and Molecular Enzymology* and *Gene Structure and Expression*.

PUBLICATION TIME

In the previous chapter, we have seen how the conflict that had arisen between the Managing Editors and the Publisher had been resolved. The Publisher had achieved his aim of rejuvenating the Board of Managing Editors, the latter had achieved their aim of a clear agreement concerning their responsibilities and also their right to be consulted in those matters where the prime responsibility clearly lay with the Publisher.

The way was now clear to tackle the recurring difficulty with scientific journals – the long publication time – and perhaps the more serious problem, the long gap that sometimes existed between the date on which an author sent the manuscript to the Editorial Secretariat and notification of acceptance or rejection. The first problem was a technical publishing one, the second was more of an administrative nature. Both were attacked vigorously by Hillier, now the respresentative of the Publisher, with notable success, as shown in Fig. 11. (The slight increase during 1984 was due to a postal strike in The Netherlands during November 1983, and temporary staffing problems in the Editorial Office during 1984.)

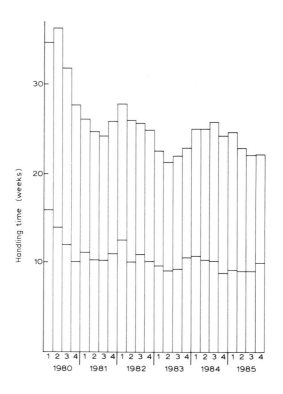

Figure 11. Handling times of Regular Papers published between 1980 and 1985, per quarter year. The upper portion represents the Publisher's handling time, the lower portion that of the Editorial Board and Secretariat.

LANGUAGE OF BBA

When BBA was founded, three languages were considered to form the international scientific languages – English, French and German (in alphabetical order) – and this was reflected in the languages that could be used in papers submitted to BBA. Although English predominated even in the early volumes, many papers were published in French and German. In October, 1981, however, a decision was taken that saddened Slater, although he accepted its logic. The minutes read:

> "In view of the small number of foreign (sic!) language papers submitted and the single-language policy of other European journals (including the new *EMBO Journal*) it was decided that only papers in English would be published."

This was stated in the 1982 Information for Contributors and was written into the contract between the Publisher and the Managing Editors.

COMPUTERIZATION

To help solve the administrative problems, the operations of the Secretariat were computerized in the early 1980's, first to assist in the choice of reviewers and, from September 1983, in the manuscript administration.

The computer, chosen to be compatible with existing hardware in the Company, is a Data General S-120 Eclipse. This has an internal memory of 512 kB, with external memory of 147 MB available on Winchester disk. A number of Data General D-200 and D-400 terminals, and three printers of letter-quality, draft-quality and hard copy are used.

The software (the UNIQUE TPS or Transaction Processing System Package) consists of an interactive screen programme, as well as background programmes written in COBOL. The system is a menu-driven transaction-oriented processing system with a number of files for on-line entry and update of data via terminals, and a general so-called retrieval function which allows a search in and between files. The interactive screen processing includes an edit mode and also checks for range errors, numeric or alphanumeric data, invalid dates and consistency. Access is via passwords, with some of the control files having limited access.

To alert the Editorial Secretariat to the most serious cause of delays, namely reviewers failing to reply, a warning list is generated of those reviewers who need reminding and those manuscripts that have been out for review for more than 4 weeks, i.e., when a decision must be made. The Secretariat can then telex or telephone the reviewer to ask for his report, or decide to consult another reviewer or a Managing Editor.

DECENTRALIZATION

In earlier chapters, we have seen that BBA was originally managed by Westenbrink and later Slater from their laboratories, assisted by a part-time secretary. When BBA exploded in the sixties, this was no longer feasible, and a BBA Secretariat was set up, housed within the Publisher, but working under the direction of Managing Editors. Gradually most authors came to realize that Jonas Daniël Meijerplein 3, Amsterdam and Vondellaan 24A, Utrecht were no longer the addresses to which papers should be submitted. After 1965, the Editorial Secretariat, with its address in Amsterdam, became an essential feature of BBA. In 1968, serious consideration was given to setting up a separate office, centred around Neuberger in London, to be run by Morris, who after doing sterling work in the early days of the Secretariat in Amsterdam, wished to return to England.

BBA was in no way a Netherlands or even an European journal (see Fig. 12). In the period 1976–1980, only 4% of the articles submitted to BBA came from the Netherlands, and 36.7% from the whole of Europe. The most papers came from U.S.A. (38.5%) and Japan (10.7%). This distribution reflects roughly the distribution of the world's biochemists. Partly in order to reflect the international character of BBA, but especially with the aim of reducing handling times, improving the selection of referees and bringing the Managing Editors in more direct contact with Editors, referees and authors, it was decided in 1981 to experiment with a more decentralized handling of papers. Although the main office was

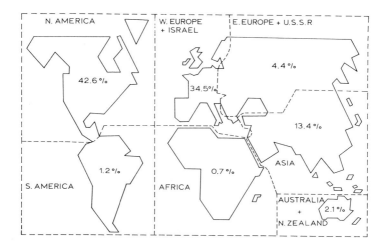

Figure 12. Geographic distribution (% of total) of papers submitted to BBA in 1976–1980. The total number submitted was 19,335.

still to be in Amsterdam, each of the Managing Editors outside the Netherlands – then P. Cohen, R.A. Flavell and G.K. Radda, all in the U.K. – would also maintain an office. In the event, Flavell moved to the U.S.A. shortly after this decision was taken and the experiment was confined to Cohen and Radda. It was originally envisaged that a computer link-up between the central office and those run by Cohen and Radda would be put into effect immediately, but in fact there were technical delays and it was not until late 1984 that the link was finally made.

During 1985, the experiment with offices associated with Cohen and Radda having been rated as successful, it was decided to set up an office in the U.S.A. associated with a newly appointed Managing Editor, Dr. E. Kennedy of Boston, Mass.

Related but nevertheless separate from the question of decentralization of the Board of Managing Editors was the question of how far the sectionalization of BBA should be reflected in the specific duties of the Managing Editors. In fact, as related in an earlier chapter, one of the alternatives considered when the journal lost Westenbrink was to decentralize the journal with one Managing Editor for each section. Slater expressed a liking for this idea in principle but did not think it feasible.

By the early 1970's, a *de facto* partial splitting had evolved, with Weissmann and Gruber dealing with Nucleic Acids and Protein Synthesis, Van Deenen with Lipids and Biomembranes, Neuberger with General Subjects, Neuberger and Gruber with Proteins, Slater with Bioenergetics and Enzymology, but there was considerable overlapping and all Managing Editors dealt with papers from all sections. This was particularly necessary during the not-infrequent absences (to attend meetings and even to take a vacation) of the busy Managing Editors.

The sectionalization was taken a step further when BBA Reviews were started, each with its own Editorial Board and Managing Editor. Cancer Reviews had, beside Weissmann, a Managing Editor outside the main Board.

The possibility of splitting the Board according to its sections was a continual topic of discussion, as is clear from the minutes of meetings on October 1, 1969, January 13, 1970, March 4, 1972 and October, 1974. Opinions were divided among the Managing Editors and it was decided to ask the opinion of members of the Board, which by that time comprised 135 members. This was done by letter of December 30, 1974 to the Editorial Board which included the following paragraph.

> "The present 135-member Board of the whole Journal is felt by some to be a rather unwieldy body and it may be more appropriate if the relevant groups of editors could be more closely associated with the policies and image of the individual sections. While it is not the intention that BBA should undergo 'mitosis', the identity of each section and of those controlling its contents could well be promoted. The aim would be to encourage editors to become more involved in journal affairs and to stimulate a feeling of cohesion between the Board members of a particular section. Symposia and conferences

could provide opportunities for Board Meetings where section policies could be discussed.

Only the names of the editors involved in the reviewing and policy of a specific section would be included in the Editorial Board printed in that section. The Secretariat have drawn up tentative Boards for each section to give you an idea of what is intended. The allocations were made by analysing the sections for which each editor was most frequently consulted in 1973; they are certainly not intended to be final. Nor is it our intention that Editors should *only* be consulted on papers submitted (or allocated) to the section with which they will be associated officially: BBA remains one journal. It is proposed that all editors should be on the Board for General Subjects and in addition must be on at least one of the specialized sections. The Associate Managing Editors would also be allocated to the sections of their speciality, while Professor Slater would remain Managing Editor for all sections of the journal.

The Managing Editors would very much like to have opinions on whether it is advisable to make this change in BBA. The advantages indicated above need to be weighed against possible disadvantages. Readers might see such a move as a step towards splitting BBA into separate journals, although this is not the intention. Such a move might also tend to formalize the compartmentalization of the journal into its present sections, and many may feel that this is not the most appropriate division of the subject. There are probably other disadvantages (and advantages) that have not come to the fore."

There was a very good response to this letter, namely 91 replies from 116 to whom the letter was sent. Of these, 42 gave unqualified support for dividing the Board, 16 qualified support, 16 were indifferent or undecided, 19 were opposed to division, 10 strongly so. Many editors commented in detail, particularly those opposed to the proposals.

In the February 1975 meeting, these replies were discussed. Slater pointed out that a number of serious objections had been raised by the Editors and he felt that these should be considered together with the general trend to de-sectionalize other journals. As none of the Managing Editors felt strongly that the Board should be divided, it was decided to maintain the present unified Board. This decision (not the analysis of the answers) was communicated to the

Table III

Distribution of papers received at different centres according to section (1984–5)

Section	Percentages of papers received in section by:		
	Amsterdam	Oxford	Dundee
Bioenergetics	9.8	8.6	3.2
Protein Structure and Molecular Enzymology	15.0	22.4	29.3
General Subjects	20.7	30.8	22.6
Lipids and Lipid Metabolism	14.8	3.1	3.2
Biomembranes	24.5	18.3	4.4
Molecular Cell Research	8.9	15.1	34.1
Gene Structure and Expression	6.3	1.7	3.2
	100.0	100.0	100.0

Editorial Board with the rider that the Managing Editors were still very keen to bring together editors most closely involved with a particular section, and would look out for suitable opportunities of putting this into effect.

This is an illustration of Slater's view that, whereas consultation is important, not all decisions need be taken by majority vote!

Thus, although the biochemical interests of its members differ, the Board of Managing Editors remains an entity and the limited decentralization that has been carried out has not led to a corresponding sectionalization of the Board. This is illustrated in Table III. Although there are significant differences in the distribution, the two new centres are not restricted largely to one or even two sections.

THE CRYSTAL BALL

For the last forty years or so, at regular intervals the imminent demise of scientific journals has been predicted. Naturally, this has also been a regular topic for discussion at the meetings of the Editorial Board with the Publisher, or at the dinners that followed them. In January, 1968, an article by Brown, Pierce and Traub [2], from the Bell Telephone Laboratories, in which a computer-based retrieval system was proposed as a replacement of journal distribution, was discussed in some detail. In the minutes of this meeting Slater is recorded as stating that he considered it a fundamental mistake to replace journals: something should be built around the journals to aid retrieval.

This is, in fact, what has developed in recent years and different organizations have made available such computer-based information-retrieval systems. One of these is *Excerpta Medica*, a publisher that fused with Elsevier (then called 'Associated Scientific Publishers') in 1970. Thus, the expertise to cope with and indeed to mould the application of the new developments in information science to scientific publisher is in house.

The earlier computer-based information retrieval systems were based on the abstract of the paper or on the 'key words' listed on the published article (in BBA since 1979). The information to be retrieved was thus restricted to that supplied by the abstractor or the author, most of whom are not experienced in providing key words suitable for such retrievals. In the 1980's, however, so-called 'natural language strategies' were introduced, whereby one can rely on the computer to match selected words anywhere in the article.

REFERENCES

1 Editorial (1982) *Biochim. Biophys. Acta* 714, 1–5
2 Brown, W.S., Pierce, J.R. and Traub, J.F. (1967) *Science* 158, 1153–9

CHAPTER 8

1986 and beyond

J. Hillier

It is difficult to consider the future of journal publishing without taking into account developments in information technology. Throughout the years of BBA's existence, technology has been used as appropriate to increase the efficiency of its publication. While much of this has been invisible to the author, referee, editor or reader, it has had a significant impact on the Publisher's office. We have seen the change from lead to computer typesetting, the introduction of computerized Secretariat functions, and many more cases where technology has been adopted.

The author, too, has seen the impact of technology in his laboratory and office. With the introduction of dedicated word processors in the early to mid seventies, and the introduction of the personal computer with word-processing software more recently, the efficiency of preparing manuscripts has been increased. However, the author still sends a paper manuscript to the journal editor or Publisher.

From the point of view of the reader, most still prefer to read a paper journal, to take with them on the train or to read at home. The convenience of the paper product is difficult to overcome with present day technology. Unfortunately the reader has to tackle the problem of information overload as well, and this is becoming increasingly difficult with paper products. Information technology can help to reduce this problem substantially. There is increasing evidence to show that the scientist is willing to use information technology to assist him with information retrieval, whether for current awareness or for reference purposes. Today, most of the information databases are bibliographic, e.g., Excerpta Medica, Biosis, and are distributed via telecommunications networks by intermediaries or 'hosts'. These hosts are usually responsible for the marketing of the information services provided. Increasingly we are seeing databases of full text of journal articles and books (without graphics), e.g., Mead with its legal database Lexis, and medical database Medis, and BRS/Saunders with its Colleague service, but these still fall short of the needs of the scientist. Currently, there are no full-text databases for the biochemist.

Technology can solve the fundamental problem of information overload for the scientist, and can significantly improve the efficiency of the publishing process. For these reasons it will be introduced, and as users see the advantages, it will be used. When this initial hurdle is overcome, the usage of 'machine readable' information will begin to take the place of the paper product.

In a recent survey carried out on behalf of Elsevier Science Publishers it was concluded that we could not expect a significant decline in sales of the printed journal, as a result of the introduction of technology, in the coming 10 years and probably not in the coming 15 years. This implies that until the end of this century we will see a duplication (and presumably increased costs) as paper product and machine-readable database stand side by side.

It is widely forecast that the scientist himself will eventually search databases to retrieve information, as opposed to the present situation where most on-line searching is done by information specialists and librarians. For this step to occur, searching languages must become simpler and the 'user friendliness' of services needs to be improved. There are currently movements in this direction with

'After Dark and Brkthro' from Colleague and 'Knowledge Index' (Dialog), but these represent only the beginnings of new development. Much more will happen as information providers and hosts obtain feedback from users concerning their wishes and as companies compete for a more favourable market position.

In considering how information technologies will impact on the publication and usage of BBA, one should look at the entire publishing process and divide it into more easily manageable segments, as follows:
- the role of the author
- the peer review system
- editorial and production
- distribution
- usage.

TECHNOLOGY AND THE AUTHOR

Recent estimates suggest that up to 80% of academic authors in the U.S.A. will produce their manuscripts using word processors or microcomputers by 1988. While our own surveys indicate that a lower proportion can be expected in other parts of the world, the trend to using word-processing technology instead of traditional typewriters is obvious. While at this time the output is only textual, i.e., graphics and complex formulae are missing, it is still in machine-readable form, usually on a floppy disc, and can be used directly by the Publisher.

In an ideal world, the text of a manuscript should only be keyed in once, at the source. Re-keying by author or typesetter is expensive and wasteful. While we are moving to this ideal, some fundamental problems exist, most notably a lack of standards which the author can follow for the mark-up and keying in of text. Without such a standard, as now, the Publisher can accept a floppy disc or tape only if he has the ability to read and display the output. With the number of different possibilities available, this is an expensive and time-consuming process. In recognition of the need for such standards, the Association of American Publishers, together with a number of other bodies, have set up a study group which will develop a Standard Generalized Mark-up Language (SGML) which, with some further specification, will become available to BBA authors, so that manuscripts can be submitted in machine-readable form using standard coding. As opposed to coding for typesetting or word-processing, the SGML represents the intellectual content of the text and not simply its physical form, i.e., it is used to identify the text elements of the BBA article, namely title, author, abstract, references, etc. Furthermore, the coding is completely neutral and non-device-specific. We expect to see the SGML introduced in the latter half of 1986 or in 1987. Authors wishing to submit manuscripts in machine-readable form need only a copy of the instructions, which could be printed in the journal, or request a pre-coded floppy disc from the Publisher which they will then be free to copy.

The above addresses only straight text, although this in itself is a significant step in the right direction to increasing the efficiency of the publishing process. The problem of half-tones, complex formulae, etc. remains. It is not expected that an economically viable solution will be found in the coming 5–10 years. While text and line drawings can be transmitted via telecommunications between author and publisher, complex graphics are both slow and expensive or even impossible. The introduction of fibre-optic telephone lines and broad-band transmission will help, and may even solve this problem, but they will not be in place in sufficient quantity for the next 10 years.

At the present time we do not perceive authors clamouring to send us their manuscripts on floppy disc, although there is a basic level of interest and it is growing. When SGML is introduced and the Publisher feels that manuscripts can be efficiently handled in this way we can expect to see the Publisher more actively promoting these developments. It should be pointed out, however, that this will not happen overnight. The Publisher also has to learn the new tricks of the trade and needs to be able to guarantee the quality for which he has a reputation.

There is still a lack of standards in the personal computer and word-processing industry, e.g., different and incompatible operating systems, floppy discs in different sizes and densities, incompatible hardware, etc. All of these add up to an increased cost to the Publisher if he is expected to handle them and it is possible that he will have to set some standards of his own in order to be able to take advantage of these developments. However, it is preferable that the industries develop their standards first to take the problems away from the author.

THE PEER REVIEW SYSTEM

In our extensive surveys of the scientific community, including biochemists, we have attempted to determine user reaction to a totally electronic publishing environment. While scientists can appreciate and probably accept machine-readable input, storage, output and distribution, the major concern is frequently expressed that to set up such an electronic publishing environment will open the flood-gates to all sorts of unrefereed information. While this is entirely feasible, we cannot envisage BBA opening itself up in this way. The peer review system is required and will therefore remain in place. There is no doubt, however, that we will see a stepwise introduction of technology into the peer review system itself.

Typically, for BBA, it takes 5–8 weeks for a manuscript to be sent to two or three referees, by the BBA Secretariat or Managing Editor, to collect the comments and give a decision to the author. Much of this time, probably at least 50%, is taken up by the manuscript and comments being transported by one mail service or another. This can be improved using today's technologies of electronic mail and facsimile transmission – unfortunately the costs of these services are still such that they act as a deterrent rather than an opportunity for Publishers. This will change as increased rates of transmission, such as 2400, 4800, 9600 baud, become more common, and facsimile machines become faster. This, of course, also requires that compatible hardware be available in the referee's or editor's office.

During 1985 the BBA Secretariat have carried out an experiment with a selected group of Editorial Board members. The purpose of the experiment was to determine the time savings which would be made by communicating with referees using electronic mail instead of regular postal services. Results from the first six months of the test showed a saving of 7.4 days from N. America and 5 days for Europe. This represents a saving on editorial handling time of 33% and 23%, respectively, or ± 10% on total publication time. Thus, savings can be substantial. If it were cost-effective or possible to transmit the entire manuscript, savings would be much greater.

Electronic mail is not in common use among biochemists, although usage is growing. Many problems were met in introducing electronic mail to our referees and many fell out of the experiment as a result. However, given the positive outcome with successful users we anticipate greater usage of electronic mail in the future.

Since the administration of the BBA Secretariat was computerized at its office in Amsterdam, we have noted enthusiasm for its use and an increase in efficiency. Our Managing Editors in the U.K. have been linked directly to the Amsterdam computer since 1984, so that they can receive manuscripts directly from authors but stay within a centralized administration. From January 1986 the computer link will be extended to Boston, U.S.A., so that U.S. authors will be able to submit their manuscript directly to a BBA Managing Editor in their own country. Depending on the success of this new venture, we will consider expanding the facility to other countries.

It will not be easy to adapt the traditional peer review system to using new technologies so that the efficiency of the system can be increased. Indeed, it may well be the last of the traditional publishing procedures to succumb, in view of the large numbers of referees and editors, the many different types of hardware and the lack of standards making it very difficult to overcome the problems. The practical problems for the editor and referee should also not be underestimated. A recent review [1] dealt with the recent experiences of a Canadian group who have tackled these problems.

EDITORIAL AND PRODUCTION

The typesetting procedure was the first part of BBA production to be computerized. This started in 1980 at Elsevier's typesetting plant, Northprint, in Meppel, The Netherlands. Thus BBA has been available (with the exception of the graphics) in machine-readable form since that time, and as such is, in theory at least, available for the manufacture of machine-readable products, whether on-line databases or others.

The goal for the production process is to automate as much as possible so that costs can be contained and the efficiency of production procedures can be increased. Thus, developments in page make-up and graphics software combined with improvements in typesetting hardware and software and printers will all help to achieve this.

In considering how technology can be used optimally to help produce the range of products which the market requires, we have had to rethink our traditional production method. So the system, which is still in use today,

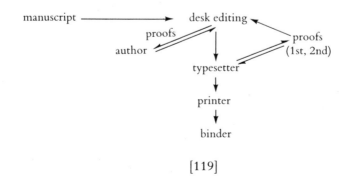

is less efficient if we consider future developments, namely, the submission of machine-readable manuscripts and the need to make machine-readable as well as paper products, as well as automating indexing systems.

Our goal is to produce a database of neutrally coded (SGML – see p. 116) data (in this regard we can consider a BBA manuscript broken down into data elements) which can then be recoded to produce the product of choice whether this be via a typesetter to make paper products, to make magnetic tapes for vendors for on-line distribution or output of citations and abstracts for secondary publishers. Such changes will have an impact on the function of the desk editor, who will become more and more involved in editing on screen as opposed to on paper, thus he will need to be trained in this new way. This process of retraining has already begun in anticipation of future developments.

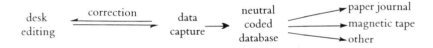

DISTRIBUTION AND ORDERING

80% of Elsevier journal subscriptions are from libraries (University, medical school, industry etc.) who use subscription agents to service their orders, claims etc. The subscription agent accumulates and sorts the orders and passes them on to the Publisher. Between Elsevier and the major subscription agents this process is now computerized, reducing cost and increasing efficiency and speed of handling orders. This process will continue as the smaller subscription agents automate their procedures and possibly one day the individual biochemist will also be able to order his subscriptions using his personal computer either via subscription agent or direct from the publisher.

The distribution of scientific information is also expected to change. While on-line databases, e.g., Excerpta Medica, have been available for many years, these have been restricted mainly to bibliographic databases which have been targeted to the library market. In the last year or so we have seen the usage of on-line distribution extended to full-text products; thus, entire journals are now available on-line via Colleague and Medis (graphics are still excluded) and these are being marketed to the end user. While at the moment the physician is the market of choice, this will expand to basic sciences, including biochemistry, in the future.

New media will also be introduced to challenge the paper product, although they will probably be distributed in the same way. These new media, in particular those that use laser disc technology, e.g. compact disc – read-only memory (CD-ROM), offer tremendous storage capacity (550 megabytes) for information as well as software which will enable the user to search the information on the disc to retrieve what he wants. Thus these discs can replace metres of back volumes of BBA and at the same time enable rapid retrieval of the information.

THE USER

As information technology is adapted, demands will be placed on the user. He will learn to interact with a (micro) computer to obtain information and he will become familiar with the computer screen for this purpose. To a large degree, biochemists are already familiar with using computers either in their experiments or for word processing or financial management. Investment in a modem will give access to on-line services and in the near future investment in a work station will enable CD-ROM information services to be used. What is not clear at the moment is how these services will be paid for or who will pay. Indications are that information retrieval is becoming a regular part of a grant request in the U.S.A. In other countries that operate different granting systems this will be more difficult to resolve.

As the market for these new products develops, the users will place more and more demands on the Publisher and the technology itself. Ease of use (user friendliness) is a primary consideration for all concerned and improvements will be demanded. It is likely, and indeed desirable, that we will move from a product- or technology-driven market to a user-driven one as biochemists (and other scients) realize the power of the microcomputer in information retrieval.

REFERENCE

1 Coward, H. and Staundera, O. (1985) *Computer Compacts* 3, 48–51.